Sir Isaac Newton

Brilliant, Bonkers & British

This book is dedicated to Max, lovely Honey and a young alchemist & physicist called Bruce.

Sir Isaac Newton Brilliant, Bonkers & British

First published in England in 2020 by AAI Ltd
1120-p
Sold in the UK, Europe, Asia, Japan, USA,
Canada, India, Brazil, Mexico, Australia.

ISBN : 979-8-619-89033-4

www.amazon.co.uk

An EASY READER book with wide-spaced lines

Isaac Newton was 5'6"
tall, along with Beethoven
and Napoleon.

quirky

/ˈkwəːki/ *adjective*

Characterised by peculiar or unexpected traits or aspects.

narcissist

[ˈnɑːsɪsɪst] noun

Person with excessive admiration of themselves, a superiority complex, lack of empathy to others, negative reaction to criticism.

genius

[ˈdʒiːnɪəs] *noun*

An exceptionally intelligent person or one with exceptional skill in a particular area of activity.

polymath

[ˈpɒlɪmaθ] *noun*

A person of wide knowledge or learning, whose knowledge spans a significant number of subjects and draws on complex bodies of knowledge to solve specific problems.

alchemist

[ˈalkəmɪst] *noun*

A person who practises alchemy, who transforms or creates something through a seemingly magical process. Origins in the Greek word khemeia, meaning "art of transmuting metals".

"*Genius*

is

patience".

Isaac Newton

Contents

Foreword

To be a scientist you have to be different. Think outside the box, make a new box. Forget the accepted and strive for the unknown. You must be bold and determined. You need to be patient and precise. To reach for the stars and travel to a place in your mind where nobody has gone before. It can be lonely when others can't see. Frustrating along the way. Bringing light where there is dark. Creating understanding where there is none. It is a challenge, a mission and a voyage of discovery, guiding our race to an enlightened and better place.

Although it's nice to be nice and good to be good, these are not traits required to be a scientist. You don't have to be fine looking or fashionable. You don't have to be considerate or tall. The person this book is about was none of these things, but he was, as a scientist must be, very, very, curious.

In the past scientists have often been portrayed as mad. Maybe because they looked mad, cared little about their appearance, focused only on the current, all-consuming experiment. Bad hair days were permanent, they appeared crumpled and disorganised. Having little or no time for anyone who couldn't understand what they were doing. Or maybe it's because they didn't think like the rest of us. They were unique, had unique thoughts, so seem detached from the masses.

Current day scientists at least appear relatively conventional, but right into the 20th century many great scientists were conducting vital, world-changing research while also, in some cases, presenting odd and peculiar behaviour. In some ways it's probably to be expected, after all it must help to be a little bit different to pursue ideas that no one else has thought of and no one else believes in yet. Many scientists have had eccentric or prickly personalities, while others are polymaths who simply can't understand the limitations of other people's feeble brains. Quite a few have gone to extraordinary lengths in their quest for knowledge, with both terrifying, violent and amusing results.

This book sheds light on one particular mad scientist who, although an extremely serious man, led a strange, surprising and sometimes difficult life. A man who brought to the world answers to enormous questions but also thought he was beyond the law, ahead of the church and specially selected by God.

Isaac Newton was a true renaissance man with unequalled accomplishments in many fields including astronomy, physics, what we now call chemistry and mathematics. Even today Newton lives on in the human psyche. We all talk like Newtonians. We might refer to the **acceleration** of industrial processes, the **force** of popular opinion or the bureaucratic **inertia** of the government. We talk about the **gravity** of a situation, the **spectrum** of choices and the **momentum** of a political party. Before Newton was born these phrases did not exist. Even those who know nothing of Newton's laws have had their minds saturated by his thoughts. To examine the work of Newton is to study your own origins.

"I do not know what I may appear to the world, but to myself I seem to have been only like a boy playing on the sea-shore, and diverting myself in now and then finding a smoother pebble or a prettier shell than ordinary, whilst the great ocean of truth lay all undiscovered before me".

Newton's Signature

NEWTON

English scientist and mathematician Isaac Newton (1642–1727), born soon after the start of the English civil war, was instrumental in the scientific revolution of the 17th century. He was certainly one of the most influential scientists in history, however Newton was also odd, eccentric and decidedly quirky. He was totally brilliant and yet dark, moody and socially awkward.

Newton emerges as an often-divisive figure, one who lived a largely solitary life. In the secrecy of his study and laboratory, we find that he also delved into heretical religion, alchemy and the occult. Even as an undergraduate he was expressing interesting, unconventional views on the soul and the end of the world. The physicist and mathematician was known to carry out ridiculous experiments on himself. During a worrying period of his life, whilst studying optics he woke up in the morning, sat on the side of his bed and contemplated for many hours. This was his morning habit, where it is said that often his mind was filled to bursting with a rush of thoughts that overwhelmed his brain. On one particular morning he decided to poke a long leather needle into his eye so far that it touched the bone behind explaining -

"betwixt my eye and the bone as near to [the] backside of my eye as I could".

It seems he did this just to see what would happen. Luckily for Newton no permanent damage occurred, although I imagine it hurt quite a lot at the time.

Another day he sat and stared at the sun for as long as he could stand it to establish how it would affect his vision. Again, he was very fortunate and suffered no long-term damage. However, he did need to remain in a darkened room for a week, before his eyes readjusted and healed.

Besides some odd beliefs and strange behaviour, Isaac had some amazingly brilliant ideas, including his work on universal gravitation and developing the Three Laws of Motion. These ideas now form some of the basic principles of modern physics. His development of calculus as a student came about in part, because he was hemmed in by the constraints of the basic, conventional mathematics of the day. So he led the way to a more powerful and entirely new technique for solving certain mathematical problems. But it took twenty-seven years to pass before he eventually decided to share his method with the world. He used calculus to aide his own personal experiments and to help explain his theories of gravity and motion. His new specialised form of mathematics was originally named by Newton as 'fluxions'. Fluxions charted the constantly changing and variable states of nature such as force and acceleration. Calculus did this in a way that existing algebra and geometry had not been able to.

Whilst he is best known for his work on gravity, Newton was a man of unending thoughts about all sorts of random things. More often with concepts than physical inventions. Although he did invent reflecting lenses for telescopes, which produced clearer images in a more compact telescope, compared with the refracting models that had reigned during that era. Looking at another area of optics he revolutionised the current thinking on light and laid out the basics for a new science called Spectroscopy. As usual he had no interest in sharing his conclusions for over 30 years.

When the Great Plague came and shuttered Cambridge in 1665, Newton went home to escape the disease. A bleak time in the history of London because of the devastating effect of the plague, but for Newton it was his "Annus Mirabilis", his Year of Miracles. This was the year he started formulating some of his best ideas. Ingenious thoughts rolled out of his mind including his theories on calculus and light and colour.

Newtons home was his family's farm named Woolsthorpe Manor. This was a place where Newton did far more thinking than farming and was also the setting for the falling apple, that is said to have inspired Newtons work on gravity.

Physicists today still don't fully understand gravity. However, Newton was the man that realised that although invisible, it must exist because some sort of force was acting on falling objects like apples, or why would they start moving from rest. He was the first to view this untouchable power as a 'force'.

"No great discovery was ever made without a bold guess".

Isaac Newton

Chapter 1

Home Life

Newton's parents Hannah Ayscough and Isaac Newton Senior married in April 1642. Newton was born an hour or two after midnight on 25th December 1642, in the manor house of Woolsthorpe. The tiny hamlet of Woolsthorpe lies close to Grantham in the county of Lincolnshire. That same year, the English Civil War commenced.

He was born on Christmas Day, according to the Gregorian calendar which was in use in England at that time. When they changed the calendar in line with our present day calendar, his new date of birth was the 4th of January 1643. The Gregorian calendar wasn't adopted by England until 1752, long after Isaac Newton's death. Original records indicate that he was baptised on New Year's Day and when the Gregorian calendar was finally adopted by England, it needed to be adjusted by 10 days, making January 4th Isaac's recognised birthday.

His birth was premature, he was sickly and as a tiny baby, with the medical facilities of the day, wasn't expected to live long. He was so weak he required a bolster around his neck to hold his fragile head up. So ominous was his situation that when two local women were sent to collect medical supplies a few miles from Woolsthorpe, they idled along the way, convinced that the baby would die before they returned.

Hannah, Newton's Mother was alone at the time as her husband, Isaac Newton Senior had died aged only 36, in the October. Just six months after her wedding and three months before the birth of her first baby. Isaac's father was an illiterate yeoman, who has been described as a wild, extravagant and an extremely weak man.

The small boy pulled through and grew stronger. Three years later his mother married for the second time. However, when his new stepfather, the 63 year old Reverend Barnabas Smith, a man almost twice his Mother's age proposed, Isaac wasn't included in the deal. Isaac remained at Woolsthorpe with his maternal Grandmother and Grandfather, who not only taught him to read and write, but raised him for many years. Sadly little Isaac didn't receive a great deal of solace from his Grandmother. He managed but they certainly weren't close. In all the writings and scribbles Isaac left behind there is not a single affectionate recollection of her.

Young Newton clearly missed his mother for eight whole years between 1645 and 1653, when she spent most of her time at Smith's rectory. During her time with Smith she produced Newtons three half-siblings, who would be Newton's closest relatives after his mother's death in 1679. Because he never knew his father, on top of the absence of his mother, Newton was left with a lingering insecurity that followed him around for the rest of life.

Isaac developed a stutter and he sometimes became frustrated when he couldn't get his words out and put his point across. However, this did not deter him from his studies or actually hold him back in any way. People who shared the same frustrating affliction include Charles Darwin, Aristotle, Moses, Winston Churchill and King George VI.

Newton understandably harboured feelings of intense dislike for his stepfather and resented his mother for pushing him aside. We know this as Newton was a meticulous list maker and one of his preserved lists included all of the sins he believed he had committed up until the age of 19. One of them included -

"Threatening my father and mother Smith to burn them and the house over them."

Following the death of Newton's Stepfather in 1653, his Mother returned to Isaac at Woolsthorpe along with her three new children, two daughters and a son from Smith. Hannah was a fairly wealthy woman by then due to her inheritance from Smith and two years after she returned to Isaac, he was enrolled at a puritan boarding school in Grantham in 1665, aged 12. His mother's brother, William Ayscough had gained an M.A. at Cambridge University many years earlier and was now the Rector of a small village called Burton Coggles, just five miles East of Woolsthorpe. As well as his religious duties he was also the Headmaster of the Grantham school. He had convinced his sister that Isaac had academic ability and that he should be prepared for study at university. This pleased Isaac and he attended his Uncle's school forthwith.

During term time Isaac lodged with an apothecary (Chemist) called Mr William Clarke whose home was near The George Inn, close to the school. Clarke liked Isaac and was impressed by the boys inventiveness and curiosity. He encouraged the lad to investigate his ideas and young Isaac learned to grind chemicals with a pestle and mortar. Clarke taught him how to measure the strength of storms by jumping into them against the wind, then comparing the distance of his leaps. He coaxed Isaac to build a small windmill created specifically to be powered by a mouse running on a treadmill and a four-wheeled cart he could sit in, power by turning a crank. He also created a kite that carried a tiny lit lantern on its tail which Isaac flew at night, worrying the locals. Clarke was probably the first person to delight in Newton's creativity and applaud him for his efforts.

Though Isaac found a friend in Clarke due to their common interests and the sheer belief Clarke had in young Newton, Isaac wasn't popular at school. He enjoyed learning new things but he didn't really fit in with his classmates.

At school he stood out as different and was clearly intellectually superior to the other boys. He was awkward and abrupt in his manner and as a result he received a negative reaction from the other pupils and didn't establish any boyhood chums.

During his school-age years he wasn't taught a great deal of maths, he disliked poetry and literature but was fascinated by technology and mechanics. As his father had been a farmer, his mother hoped that Isaac would follow his father and one day also be a farmer, but Isaac disliked farming. He preferred to develop sundials, which turned out to be extremely accurate.

Regardless of Isaac's preferences, at some point in 1659 Hannah pursued her own ambitions for Isaac's future. She deemed that her eldest son would take on and manage her recently inherited farm property and withdrew him from school, aged 17. It wasn't long before Hannah realised that this was not going to be a long-term solution. In fact, she could see that this would be disastrous for her estate and also for Isaac. He was not able to focus on rural affairs and was totally out of his depth with cows and land.

Fortunately his Mother eventually backed down, listened to her brothers reasoning and in 1660 Newton was returned to his Grammar School, where he continued to prepare for an academic future at University.

After completing his studies at Grantham his report noted his mechanical ability. He was commended for his skill in building models of machines, such as windmills and clocks. It was said that he had acquired a good understanding of Latin, but Mathematics was not really mentioned.

Raised, in effect as an only child, Newton was used to being alone or in the company of adults, some of which had inspired and encouraged innovation in the boy. But he spent much of his childhood tormented by loneliness which could only be alleviated by earnest focus on his childhood experimental projects and model building schemes. As an adult this pattern of sadness, isolation and extreme focus would continue for most of his life.

Regardless of all he suffered, by June 1661 he was ready to join some of the countries great minds and entered Trinity

College, Cambridge. As a result of the time he spent out of school trying to be a farmer, his start at Trinity was delayed and he was a few years older than the other undergraduates. Regardless, he had arrived at the place that would be the starting block of his amazing career and by the beginning of 1664, he had started to teach himself mathematics, taking notes on works by Wallis, Oughtred, Descartes and Viète.

Newton had help both getting to and during his time at university. As well as his uncle's assistance, in 1661 the brother of Newtons Mothers best friend, Humphrey Babington, was made rector of Boothby Pagnell. Another small village just over six miles away from Woolsthorpe. Babington had been reinstituted as a fellow at Trinity and although little documented information is known about their relationship, what is known is that Newton was close to Babington during his early years at Cambridge. It is assumed that he acted as Newtons patron during this period. These two men paved the way for Newtons entrance to the academic world of university and made his passage of education easier.

Constantly in the background of Newton's childhood was a

country in crisis. England had been gripped by public unrest which evolved into the start of The English Civil War in 1642. Charles the King of England was beheaded in 1649, making way for Oliver Cromwell to rule as Lord Protector from 1653 until his death in 1658. Cromwell's son Richard held his seat from 1658 to 1659, which was followed by the restoration of the monarchy which placed Charles II on the throne in 1660. This disturbing period must have impacted upon Isaac's family in a negative way as the whole country was in turmoil. However it brought about a positive future for Isaac via the effect it had on the universities at that time, including Cambridge. Until this point the universities had been controlled by the Anglican Catholic Church and now they had been extricated from its grip. This made it possible for Scientist Robert Boyle to seek support from Charles II and in 1660 the 'Royal Society of London' emerged as an institution. This opened up a whole new world of scientific possibilities, making the intellectual environment of Isaac's time at university radically different from when he was born.

In spite of the help he received, life was not always easy for

Newton at Cambridge. Upon the instruction of his wealthy widowed, Mother, Newton was harshly enrolled as a sub-sizer, which was an undergraduate at Cambridge who did not have to pay tuition fees. Instead they helped in the kitchen or did other household chores for their college, to help pay their bills. Other wealthier students regarded sub-sizers as servants, that would wait on their tables and clean their rooms.

During this period, as always, Isaac liked to keep a journal of his ideas and thoughts. He had written in his college notebooks about himself and his sins in which he confesses to "Making pies on Sunday night" and "punching my sister". Which shows he had thoughts like any normal young man with conventional frustrations, habits, desires and an element of mischief.

Other important evidence that helps with the understanding of Newton's development during adolescence and early adulthood, is supplied by the lists of expenses he kept from 1659-69. These notes show that the image of an unsmiling, self-absorbed, puritan Newton was not always so. It reveals that as an undergraduate he did get out once in a while, to the tavern, the bowling green and he even occasionally played cards, although it seems he usually lost. Perhaps even more surprising, is that he appears to have run an informal money-lending operation for fellow students at Cambridge, though it is not known whether he charged interest on his loans, or not.

By late 1666 Newton had become the leading mathematician of the world. Having lengthened his initial basic work and validating his development of calculus, as presented in his great publication which he bestowed upon the world in October 1666. During the course of the following fifteen years as Lucasian Professor of Mathematics, Newton presented many lectures and continued research in a variety of interesting fields.

In the latter years of his life, Newton became one of the most famous men in Europe. Regarded as a leading figure in the world of science in London and beyond. For the rest of his life he had face-to-face contact with many individuals of great importance and power, in ways that he had never experienced during his years at Cambridge.

As he grew older and matured, living on into his sixties and seventies, the world moved on and things began to change. In the Summer of 1714 Queen Anne died and the Stuart Dynasty came to an end.

During his final years Newton's health began to deteriorate, suffering from digestive problems which made him drastically alter his diet and caused him mobility issues. He lived at Cranbury Park, near Winchester and then in London with his niece Catherine Barton, the daughter of his half-sister Hannah. Teenage Catherine was a bright and excitable girl who swept the cobwebs away and brightened Newtons everyday home life. Catherine was a prominent socialite of the day who mixed with the rich and powerful. Eventually she married John Conduitt, one of the wealthiest men in London in 1717. She remained in close contact with Isaac for the rest of his life and it is thanks to both Catherine and her husband that many of Newtons written records and papers survived.

In 1722 Newton suffered an 'attack of the stone' and on March 2nd 1727 he attended the Royal Society for the last time. Sixteen days later he was suffering from severe discomfort in his abdomen, he collapsed, blacked out with the pain and never regained consciousness. This most extraordinary man finally died in his sleep between 1am and 2am in Kensington, London on 20 March 1727.

In the 1970s, long after his death, a lock of Newton's hair was examined and found to contain high levels of mercury, some forty times higher than normal levels. This was probably a result of his alchemical pursuits. Mercury poisoning is famous for sending people 'mad as hatters', because hatters used mercury whilst crafting their headwear. Mad Hatters disease, can result in insanity, but only after being exposed to it for a long time, weeks to months, for early symptoms to show. Mercury poisoning could explain Newton's eccentricity in late life but it doesn't explain his eccentric activities in his early life. It's possible that people look for reasons why the great man was not as sane as we'd like. How can a man of such genius also believe with great vigour, some of the random and occasionally bizarre things that he did. After all, although he was almost completely grey by the age of 30, Newton never actually lost all of his hair. Nor did he suffer from bleeding gums. Hair loss and bleeding gums are the two of the main symptoms of mercury poisoning.

Isaac Newton was never married and some say he died a virgin, although this has never been confirmed. However he definitely didn't father any children. As an adult Newton immersed himself in scientific work. He had no hobbies, just an everlasting curiosity and devotion to finding things out. As a result he addressed and solved conundrums that had perplexed the human race for centuries. By incorporating his mathematical genius he provided a new world view of the Universe.

Although he had been a dark, brooding figure in the scientific fraternity, disliked by many, he was also capable of great generosity and kindness. As Newton aged he mellowed and accolades of his amiability were plentiful. At the time of Newtons death he was a famous and wealthy man, who was genuinely mourned by an adoring nation. On March 28th his body lay in state at Westminster Abbey and on April 4th the Lord Chancellor was one of his pallbearers. Newton was laid to rest in the famed 10th century abbey, which also houses the remains of Monarchs such as Elizabeth I and Charles II. His elaborate tomb, paid for by John Conduitt, stands in the abbey's nave and features a sculpture by J.M. Rysbrack and W. Kent. Immortalising a reclining and thoughtful looking Newton with one arm resting on a selection of his great printed works. Charles Darwin and other noted scientists have since been buried in the same part of the Abbey, close to Newton.

The Latin inscription on Newton's tomb translates as –Here is buried Isaac Newton, Knight, who by a strength of mind almost divine and mathematical principles peculiarly his own, explored the course and figures of the planets, the paths of comets, the tides of the sea, the dissimilarities in rays of light, and what no other scholar has previously imagined, the properties of the colours thus produced. Diligent, sagacious and faithful, in his expositions of nature, antiquity and the holy Scriptures, he vindicated by his philosophy the majesty of God mighty and good and expressed the simplicity of the Gospel in his manners. Mortals rejoice that there has existed such and so great an ornament of the human race!

His monument is made from white and grey marble and was completed and presented to the world in 1731. Its base supports a sarcophagus with a relief panel. The relief depicts cherubs using instruments related to Newton's mathematical and optical work. Childlike seraphs toying with a prism and a telescope, whilst another uses a steelyard to balance the planets with the sun. Others panels depict Newton's activities as Master of the Mint, the figures carry pots of coins and an ingot of metal is being placed into a furnace.

The inscription exalts him to the highest status a human being can gain. Not just the bones of a man, but the greatest man who ever lived.

It can be said that Newton died a true polymath. A master of many subjects including mathematics, astronomy, physics, chemistry and theology. His never ending curiosity led him to examine and try to find solutions for problems as everyday as a door scratching cat and as far reaching as humanity's ultimate purpose in the cosmos.

Following Newtons death, many of his papers were deemed "unfit to publish" and remained out of public view until 1936, when Sotheby's auction house acquired and sold most of them to the economist John Maynard Keynes. These included papers about the Philosopher's Stone, which Newton was convinced could turn lead into gold and possibly be an elixir of life, along with his prediction about the end of the world.

"As a blind man has no idea of colours, so have we no idea of the manner by which the all-wise God perceives and understands all things".

Isaac Newton

Chapter 2

What Newton Got Right

In the progressive atmosphere of 17th Century England, the expansion of the British Empire made rapid gains and maturing universities like Oxford and Cambridge were producing many great scientists and mathematicians. However the one that eventually and indisputably stood out above all the rest, was Isaac Newton.

Newton studied at the University of Cambridge, Trinity College, from 1661 to 1665, when he completed his bachelor's degree. Then he attended Kings College from 1667 to 1669. The two years between courses turned out to be very productive for Newton. He wanted move straight from Trinity to Kings College to continue with his studies, but an epidemic of the bubonic plague quickly changed his plans. The university closed its doors not long after the disease had commenced its deadly rampage through London and kept them closed whilst it passed. It was a difficult and worrying time for England. During the first seven months of the outbreak, it was gauged that over 100,000 London residents had died.

Newton made good use of his unexpected break. Over the next two amazing years 1665 to 1667, he returned home to Woolsthorpe to escape the disease in London and the south. During his time at home Newton worked on his new theory of light, Invented infinitesimal calculus producing a brand new view of mathematics, as well as observing and establishing an understanding of gravity. Newton called 1666 his *annus mirabilis*, or year of miracle's.

English born John Wallis and Isaac Barrow had previously worked on calculus, but Newton had taken the idea and actually made it work. Providing a logical base for dynamic change and motion to be understood, which the ancient Greeks had not been able to fathom. Making sense of the movement of fluids, the planets and their orbits around the sun. This work was later recorded in Newton's three part book, Philosophia Naturalis Principia Mathematica.

Calculus (1665) Published in Principia 1685

In 1665 aged just 23, Newton worked to develop a mathematical theory which we now know as calculus. It is the mathematical study of constant change, which provided our race with a robust fresh method for evaluating the incline of curves and the area below the curves. At the time Newton named it his "Method of Fluxions and Fluent's". These days, it has many uses in engineering, science plus economics and can resolve many problems that algebra alone cannot. He did not publish it at the time and only decades later as a minor annotation in the back of his book the Principia.

Isaac Barrow, a brilliant man and the first Professor of Mathematics at Cambridge, had first introduced Newton to the basic idea of calculus because Newton had difficulty understanding existing mathematics from the books he bought for his studies. The geometry and algebra that was in use at that time didn't provide him with solutions to problems he wished to investigate. He was only in his early twenties and yet he had already reached the limits of current human thinking about maths. So he set about finding new techniques to address the problems he, and many others wanted to solve.

Throughout all of his research, Newton did not approach his work by what was then known as the method of hypotheses. This method, still used by many university students today, involves putting forward a hypotheses that reaches further than all known phenomena and then setting about proving the theory by means of research, testing and drawing observable conclusions from the outcome. Instead Newton insisted on having actual phenomena dictate each aspect of a theory. With the aim of limiting the initial aspect of the theory as much as he could, to allow him to focus on and draw conclusions based only on evidence and logical reasoning, from the specific phenomena he was currently interested in.

Mathematicians in the 1600's were already able to calculate the speed of a ship, but they didn't know how to calculate the rate at which the ship was accelerating. They had fathomed how to measure the angle of an airborne, sailing cannonball, but they had no method to calculate which angle would send the cannonball the furthest. What they really needed was a mathematical means to precisely calculate problems that involved ever changing variables.

Scientists of the day agreed that it was easy enough to represent and calculate the average slope of a curve, such as the increasing speed of an object on a time-distance graph. However as the slope of the curve was constantly varying, there was no method to give the exact slope at any one individual point on the curve.

By taking existing algebra and trigonometry, Newton took mathematics one more step further by adding limits to his functions. He calculated a derivative function $f'(x)$ which gives the slope at any point of a function $f(x)$. This process of calculating the slope or derivative of a curve or function is now called differential calculus. He called the instantaneous rate of change at an exact point on a curve the 'Fluxion', and the changing values of x and y the 'Fluent's'.

The area under the curve can be approximated by summing the area of a number of rectangles drawn within it. The greater the number of rectangles used, the thinner they become and the more accurate the result.

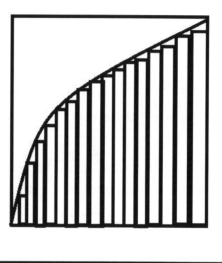

Newtons Fluxions & Fluent's (Calculus) F(X)=√ x

Basically, integration is founded on a limiting process which approximates the area of a curvilinear region by breaking it into infinitesimally thin, vertical slabs or columns.

Looking at the diagram and in hindsight, it all seems fairly clear and an obvious solution, but until that point in time nobody had actually come up with this idea. An idea that changed mathematics forever.

Having concluded the derivative function for a particular curve, it was then an easy matter for Newton to calculate the slope at any particular point on that curve, just by inserting a value for x. For example, in the case of a time-distance graph, the slope represents the speed of the object at an exact point.

The opposite of differentiation is integration or integral calculus or, as Newton's called it, the method of Fluent's. Together differentiation and integration are the two main operations of calculus. Newton's Fundamental Theorem of Calculus states that differentiation and integration are inverse operations, therefore if a function is first integrated and then differentiated (or vice versa), the original function is retrieved.

One of Newtons reasons for not publishing his new thoughts on mathematics at that time, was that he was concerned he might well be mocked about his unconventional ideas and pacified himself by sharing his thoughts only with his fellow colleagues. Besides, he had other things to think about such as alchemy and philosophy. By 1671 he had extended and completed the majority of his now lengthy written account about calculus, only to find that initially no one would publish it. This was quite a blow for Newton and he once again diverted his focus away from calculus, for over a decade.

The mathematical lectures he gave following this setback, on the whole related to algebra. He later came to regret not pushing for publication sooner when in 1684, his German peer Gottfried Leibniz published his own independent version of the theory. Newton stewed for quite some time until he eventually published his version in the Principia in 1687. The Royal Society after difficult consideration, gave credit for the first discovery to Newton and credit for the first publication to Leibniz. There was embarrassment all round when it was made public that the Royal Society's accusation of plagiarism against Leibniz, was actually authored by Newton himself. Creating difficult debate and controversy that would tarnish this discovery for Newton, for many years.

The Reflecting Telescope (1668)

Newton's Reflecting Telescope

Newton was born into an era of bulky, uninspiring, basic telescopes. The best models available used a set of glass lenses to magnify an image. Whilst still an undergraduate Newton had read works on optics and light by an Oxford student named Robert Boyle (1627-1691). Boyle was a brilliant pioneer in Chemistry and Boyle's assistant Robert Hooke (1635-1703), was an impressive experimenter and between them they had produced some interesting finds.

Newton had been inspired by these two men, although he begrudged them any credit for anything they did. Regardless, they had shown him a start that interested him and when something interested Newton he was on a mission. Very quickly he was grinding glass, performing experiments and working out ways to improve the telescope.

Through his previous experiments with colour, Newton knew that lenses refracted different colours at different angles, which in turn created blurry images for the user. Fascinated with the planets and stars, he needed to find a way to improve what was currently available to give him a better view and understanding of the solar system. To improve the current telescope Newton incorporated reflecting mirrors rather than refracting lenses. He used a large mirror to capture the image, then a smaller mirror to bounce it into the viewer's eye. His new device produced a vastly clearer images and also for the first time, made the telescope portable due to its reduced size.

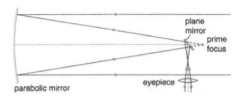

Inner Workings of the Newtonian Telescope

In 1668, he introduced his latest wonder to the world after he named it the 'Newtonian telescope'. Up until this point in history a telescope was a large and difficult to move instrument. Now people could climb to the top of a mountain with their telescope and view the stars from the best vantage points. Newton's telescope proved to be significantly more powerful than traditional telescopes and his simple lightweight design made it very desirable in its day.

Orbital Astronomy (1679)

Despite Newtons advances with calculus and his telescope, he showed only a limited interest in orbital astronomy until fellow scientist Robert Hooke wrote to him in November 1679, hoping to extract new scientific findings for the Royal Society. Newton had only recently returned to Cambridge following his mother's death earlier that year. Hooke bestowed Newton with a number of queries, including asking him what his thoughts were, relating to Kepler's laws and the trajectory of a body, under an inverse-square central force. That is to say, the square of the time it takes for a planet to complete one orbit of the sun, is proportional to the cube of its average distance from the sun. Hooke thought that maybe Kepler's laws contained something more. He and his colleagues considered that Kepler's laws fell true if an extra dimension was added. Possibly by assuming that the sun actually pulled all the planets toward it, with a force that grows weaker in proportion to the square of the planet's distance.

Newton had already discovered this systematic relationship between conic-section trajectories and their inverse-square central forces at an earlier date, but as was his way, he kept it to himself. He did not further explore this line of discovery until Halley (as in Halley's comet) turned up to visit him in the summer of 1684 and asked him the very same question as Hooke. He stared straight at Halley and replied, "an ellipse". He looked for but was unable to find his notes on the matter, but said he would forward a rewrite of his theory to Halley in London. Newton dutifully kept to his word and the following

November he sent to Halley a nine page manuscript, which he called 'De Motu Corporum in Gyrum' (On the Motion of Bodies in Orbit). In this document Newton refered to elliptical orbits, which are a derivation of Kepler's laws of planetary motion. Newton described how the Earth's orbits of the sun are not circular but elliptical orbits. He went on to explain that if an object like a planet, maintains an elliptical path around another object, there must be a force attracting it. He assumed the orbiting body would normally continue in a straight line if left to its own devices and drift of its curved orbit and continue straight ahead into the darkness of space. However if that planet had an attraction to the orbit's centre, it would be pulled back onto its orbit. Thus the constant drift and pull and drift and pull produces orbital motion.

Newton was improving on Galileo's Law of Inertia which states 'that all bodies tend to continue moving in a straight line unless acted upon by an external cause, or force'. Newton understood that by adding a force that attracts a body back toward the centre of the orbit, you have the cause of the radial motion. This was the missing ingredient that Hooke and Halley had been looking for.

To describe that mathematically and make a connection between the mathematical form of the inverse square law and the mathematical properties of orbits found by Kepler, you need to visualise dividing time into really small intervals. For each interval imagine the orbiting object as moving tangentially by a tiny amount and simultaneously, radially by a tiny amount. The total effect of these two motions returns the object to its orbit, but slightly further along it's circle than where it began. Repeating this over and over produces a ragged circular orbit. By making the time intervals small enough, the path becomes almost a circle. As the segments of the curve approaches zero in size (i.e. an infinitesimal change in x), the path is a perfect circle, or in this case an ellipse.

Calculus, Newton's new form of mathematics had made it possible for him to calculate and show to others how an orbiting body moves tangentially, then falls radially, forming a ragged path. Then he made the straight line portions of the jags vanishingly small, to smooth the path and form an exact circle. Newton's view of Orbital motion is simply the motion

and thanks to Newton we know that -

The more mass an object has, the bigger the force of attraction and the closer something is, the bigger the force of attraction. As the Earth has a very large mass and being very close to the moon, it exerts a strong pull on anything in its vicinity, including the Moon.

It was Aristotle who lived from 384 to 322 BC who initiated thoughts about gravity. Originally he believed that heavier objects fall faster than lighter objects. In fact, most people believed this until 1638, when Italian astronomer Galileo Galilei proved that objects fall at the same speed whatever their weight. It is said that Galileo proved his theory by dropping different-sized balls from the Leaning Tower of Pisa. He found that the speed at which an object falls is determined not by how heavy it is, but by how long it has been falling.

Newton calculated, at the latitude of Paris, a body dropped from rest would fall 15 feet and one-eighth of an inch in a second. To prove this he conducted experiments with a variety of materials including gold, silver, glass, lead, salt, sand, wheat, water and wood. He concluded that regardless of composition and whether on Earth or in the heavens the attraction always follows the same law.

In 1971 astronaut David Scott performed a similar experiment to Galileo's. Not in Italy but on the Moon. Scott dropped a feather and a hammer at exactly the same time and they both reached the surface of the moon at exactly the same time.

Galileo calculated the rate at which objects accelerate while falling to Earth. That rate is now known to be 9.81 m/s² (this figure alters slightly depending on whereabouts on Earth it is measured and also disregards the effects of air resistance and other factors which could prevent a frictionless fall). Galileo's work made people think about gravity, but he couldn't explain why it was that things fell to the ground in the first place.

Like most scientists, Newton used the work of previous scientists as a basis for his own theories. He looked at Galileo's findings and also German astronomer Johannes

Kepler's ideas. Kepler had worked out the rotation of the planets round the sun. To do this, he in turn had been using observations made by Danish astronomer Tycho Brahe.

Using Kepler's laws of planetary motion, Newton realised that gravity was a force of attraction and the size of the force depended both on the mass of the objects involved, and the distance they were from each other.

Newton's law of universal gravitation, states that-

Any two bodies in the universe attract each other with a force that is directly proportional to the product of their mass and inversely proportional to the square of the distance between them.

The law can be written as an equation:

$$F = G \left((m1 \times m2) / r^2 \right).$$

The force (F) between two objects of mass m1 and m2 is equal to the product of their mass, divided by the square of the distance (r) between them. G is the gravitational constant which remains the same wherever it is applied in the Universe.

Newton's law shows why two objects of different mass fall

with the same acceleration. Basically, if the mass of an object is doubled, the gravitational force is doubled, but the rate of acceleration remains the same. If the mass of an object is halved, then the gravitational force is halved, but again the rate of acceleration remains the same.

Using Newton's Law of Universal Gravitation, scientists were able to calculate the presence of Neptune before it was officially discovered. In the 1820s it was established that Uranus, the most distant planet discovered to date, was not where Newton's laws dictated it ought to be. Over the next few years astronomers in England and France independently concluded that, if Newton was correct, the movement of Uranus must be affected by another unknown planet. They then calculated the location where this other planet should be. The planet they were trying to find was Neptune, which was finally located in 1846. This proved once again that Newson's law was correct and that It was Neptune's gravitational pull that was affecting the orbit of Uranus.

Newton realised that every particle of matter is attracted to every other particle in the Solar System, but the force exerted between individual particles is very small, because their mass is very small. With larger objects, such as the Moon and the Earth, the force of attraction is bigger and stronger. So strong that it keeps the Moon spinning around the Earth, instead of drifting off into space.

The mass of the Earth is approximately 6 million, million, million, million kg. As the Moon is smaller and less dense than Earth, it has a significantly smaller mass. Therefore the force of gravity on the Moon is much smaller, only about 17% of the Earth's. So Neil Armstrong's weight on the Moon was only 17% of what it was on Earth. This enabled him to jump higher than he could on Earth, but the Moons gravity was still strong enough to bring him back down to the Moon's surface. Although the mass of the Moon is much less than that of the Earth, it still affects it. The force it exerts as it orbits the Earth pulls the oceans towards it, causing tides.

Newton had combined his laws to explain the theory of uniform circular motion and Kepler's laws of planetary motion. He highlighted the relationship between the inverse-square and Kepler's rule regarding the square of the planetary periods to the cube of their mean distance from the Sun. Thus Newton removed the last doubts about the validity of the heliocentric model of the Solar System. Heliocentrism is the astronomical model in which the Earth and planets revolve around the Sun, which is at the centre of the Solar System. Newton had proved that Aristarchus of Samos (310 BC – 210 BC) an astute ancient Greek astronomer and mathematician, who first proposed the idea of the Sun being at the centre of the Solar System, was correct.

Aristarchus identified the "central fire" as the Sun and he put the other planets in their correct order of distance around the Sun. His astronomical notions were rejected in favour of the geocentric theories of Aristotle and Ptolemy. Geocentrism places the Earth at the centre and the Sun revolving around it. This was still the main belief held of scientists for a thousand years after Aristarchus had died, until Newton put them right.

It seems that the unique idea that gravity is universal revealed itself to Newton gradually. It dawned on him as he edited his early version of the Principia. Prior to this scientists had thought that if planets did impose a force of gravity, it only affected its moons. They assumed that each planet was self-contained with its own laws. Newton had originally set out investigating a notion that the cause of objects falling down to Earth could also shed light on the Earth's pull on the Moon. He hadn't considered the pull of the Sun on the planets. When this did occur to him he contacted an English astronomer, asking for information about the comets of 1680 and 1684 and also the orbital velocities of Jupiter and Saturn, as they headed towards each other. Using these statistics he applied many complex mathematical calculations and made comparisons of his findings. Eventually he concluded that the same law of gravity applies to the entire Solar System. He re-edited the Principia once more to include this ground breaking knowledge.

Halley helped to edit the Principia for Newton and set about getting it published, as Newton had a tendency to drag his feet about these things. Halley thought it a masterful piece of work and wanted to share Newton's genius with the world. So he distributed copies of the Principia to as many of the top scientists and leading philosophers of the day as he could, with the initial small publication.

The primary press-run of approximately 300 copies of the first edition, was not large enough to be read by anyone other than Halley's select elite and unable to reach the masses. The second edition was more widely available because of two pirated Amsterdam editions. The third edition was translated in both English and later French, making it considerably more accessible. The book was a great British triumph and news of its brilliance spread quickly throughout the academic groups and intellectual meeting rooms of Europe. It was obvious to all that Newton had produced a book that would alter the way people thought about science forever.

The Principia seemed to provoke alternate understanding and meaning to different people, as it is a long, heavy read and not easily understood by the average person outside the scientific field of physics. Even those who actually managed to acquire a copy would have found much of it difficult to grasp. Most will have read portions of his writings, but few would have been able to take in the full complexity of what Newton presented them with.

In 2016 an auction house sold a first edition of Isaac Newton's 1687 masterpiece, Principia Mathematica, for just over $3.7 million. Making it the most expensive science book sold to date.

Opticks – (Optics) The Heterogeneity of Light (1704)

In the late 1660s and early 1670s Isaac Newton famously discovered that white light is a mix of colours, which can be separated into component parts with a prism. He also showed that the multi-coloured spectrum produced by a prism could be recomposed into white light by a lens and a second upturned prism. As a result Newton was able to prove that the belief of the day, that light was simple and homogeneous (consisting of parts all of the same kind) was incorrect. He established that light is complex and heterogeneous (diverse in character or content). The heterogeneity of light has since been the foundation of Physical Optics.

Newton was the first to understand rainbows, by refracting

white light and resolving it into its component colours: red, orange, yellow, green, blue and violet. Even though other scientists had been experimenting with prisms in relation to colour long before Newton, their belief was that somehow it was the prism, that coloured the light. Newton obtained two prisms and set up an experiment, so that a single spot of sunlight fell onto it. Usually in such experiments a screen was put close to the other side of the prism and the spot of light came out as a mixture of colour. Newton realised that to get a proper spectrum you needed to move the screen a lot further away.

Newton knew that as the light enters a prism from the left, it is refracted by the glass. He realised that the violet rays bend more than the yellow and red, so the colours separated. After moving the screen 22 feet across the room and achieving a beautiful spectrum, he carried out his crucial experiment to prove that the prism was not colouring the light. He put a screen with a slit cut in it, in the way of his spectrum which, as he had predicted, only allowed the pure green light to pass through.

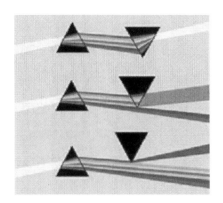

The diffraction of light

The diagram shows how Isaac Newton's crucial experiments worked between 1666 and 1672. A ray of light is divided into its constituent colours by the first prism and the resulting bundle of coloured rays is then reconstituted into white light, by the second upturned prism. Proving once and for all that it wasn't the prism colouring the light, as the second prism did not affect the uncontaminated green light. If it had been the prism that was colouring the light, the green would have emerged as a different colour.

Newton had shown the other scientists that they were wrong and he named the multi-coloured band of light the 'Colour Spectrum'. He had bypassed the scientific sceptics of the day, who thought that rainbows formed as light was refracted and reflected in raindrops, but confessed that they didn't understand why rainbows were so colourful. Prior to Newtons interest in colour the majority of scientists believed that it was the rain water that mysteriously dyed the rays of the Sun, to produce the range of colours.

From the prism experiments Newton had already carried out, he was able to develop his Theory of Colour which is thought to be his most important contribution to optics. His colour theory states that colour is the result of objects interacting with already coloured light and not generating the colour by themselves. Individual rays of light excite the sensation of colour when they strike the retina of the eye. Therefore, it was only after Newton's study of optics that light was identified as the source of the sensation of colour.

The study of orbital mechanics and optics are now regarded as subjects covered by physics. In 1704 he wrote a book on his findings about the refraction of light. He called the book 'Opticks' and his work changed the way people thought about light and colour.

Chapter 3

Portraits of Newton

In a time before iPhones, selfies and even cameras, the only way to present yourself to the world was via a painted portrait. Newton enjoyed his painted portraits and actually sat for twelve portraits, by various artists during his lifetime. A number on par with Kings and Queens of the day.

Isaac Newton as a boy. Artist unknown.

1689 – Portrait of Isaac Newton by English artist
Sir Godfrey Kneller

One of the first official portraits of Newton, seen here at the age of 47, is currently owned by the 10th Earl of Portsmouth. It shows Newton at the height of his scientific acumen, before he went to London to take charge of the Queens Royal Mint.

Death mask of Sir Isaac Newton

by John Michael Rysbrack, created in 1727

A full face death mask of Sir Isaac Newton. The mask is one of a number prepared shortly after Newton's death, by the artist Michael Rysbrack. This likeness was later used by Rysbrack to sculpt Newton's finer features in marble, for his tomb at Westminster Abbey.

*"In the absence of any
other proof, the thumb
alone would convince me
of God's existence."*

Isaac Newton

Chapter 4

How the Man Developed

At some point in his late teens Newton came to know Isaac Barrow, the first Lucasian Professor and later, master of his college at Cambridge. The Lucasian Chair of Mathematics was a position initially conceived by Henry Lucas, a Member of Parliament for Cambridge University. A prestigious role that in 1669, aged just 26 years old, Newton would be the next and only the second to claim it.

Newton's intellectual activities as an undergraduate started out as almost entirely extra-curricular. He just didn't seem interested in the subject he was supposed to be studying, namely, The Ethics and Natural Philosophy of Aristotle. His nonchalant attitude and disregard for University demands gave him an early reputation as a poor student, with little to offer. It took some time for his genius to be recognised, which arrived initially from his mathematics professor Isaac Barrow. This man became an excellent ally and a strong believer in Newtons ideas and abilities.

Upon his return to Trinity College during the spring of 1667, Newton devoted himself to two very different fields of study. He threw himself into the understanding of optics and mathematics, with particular interest in algebra. His focus on algebra paid him dividends and he was soon regarded as exceptional by his fellow graduates and the lecturers in the Cambridge Mathematics Department. He so impressed the extremely influential Isaac Barrow, that when Barrow resigned his prestigious position as Lucasian professor of Mathematics, he insisted that Newton must be the one to take over the role. The job came with a hefty salary of £100 per annum, which was a great deal of money in the 1600's and ten times what Newton's Mother had granted to him. He now had money, credibility and a fair amount of power within the University walls and he was well on his way to becoming famous.

Once he had settled down at Trinity College, he undertook an extraordinary programme of creative theological research. Work that displayed great expansiveness, originality and radicalism, totally unmatched by his peers. This programme of study was predominant and unrelenting. Only occasionally interrupted by his work in alchemy, mathematics and natural philosophy. His polished writings on theology were not the wittering of an immature mind, but the product of a sharp, committed, brilliant and courageous analyst and Newton was being noticed, all over the world.

Newtons Personal Problems

In spite of having the world at his feet, Newton suffered many difficulties along the way. Although he profited greatly from spasmodic interactions with fellow scientists dotted around Britain and on the Continent, usually by mail, it seems he was a reluctant traveller. He never ventured outside the vicinity of a compact triangle which linked his home at Woolsthorpe, to his University at Cambridge and the capital City of London. Newton never travelled abroad, not once. He studied such enormous worldly subjects, read texts from lands and cultures far and wide but chose to live in a small, insular environment.

Newton rarely saw the use of having people in his life, undoubtedly because his mother rarely seemed to have much use for him. The life of this reclusive man did not include many real friends or family that he could genuinely feel close to, not even a single long-term lover that we can be certain about. Right up until his later years, when he mellowed slightly, Newton just didn't seem to socialise for the sake of socialising. He only ever gathered in any kind of group to discuss scientific advances, alchemist findings or to attend religious meetings. A remark made by Humphry Newton, a distant relative who assisted Newton in his laboratory for five years, shows the extent of Newtons solemnity. He said that he only saw Newton laugh once during the time that he worked for him and that was when someone asked him why anyone would want to study Euclid. Newton was a serious and focused man who saw no purpose in life for frivolity.

During his lifetime he achieved a great deal of fame but didn't have anyone to really share it with. He strove for and gained intellectual recognition but it is quite possible that he didn't experience the feeling of falling in love. He was bestowed with a Knighthood and honours galore, but wasted much of his time in intellectual dispute, bickering over what was and wasn't his work. He was an arrogant, rude and generally unpleasant man to spend time with. It was only his intellect that made him so attractive to others. Most people wanted to know what he knew, but they didn't actually want to know him.

Newton found it difficult to accept criticism from fellow scientists and mathematicians and made lifelong enemies with those who dared to question his work. He famously had several angry showdowns with Robert Hooke, a well-known scientist at the time, best known for his microscopic experiments.

These two men had a long-lasting grudge match which started when Hooke declared that Newton's Theory of Light was flawed and denounced the physicist's work. Newton in turn denounced Hooke's work in the late 1600's, following Newtons experiments with his celebrated phenomenon of colours. It was during the period when people still thought that colour was a mixture of light and dark and that it was the prisms that actually coloured the light. Hooke was a believer of this established theory of colour. He had his own scale of colours that went from brilliant red, which was pure white light with the least amount of darkness added, to a dull blue, the stage before the total elimination of any light, causing true black. Newton correctly realised this theory was incorrect and wasted no time in telling the world of Hooke's blunder. Hooke was a hard and abrasive man, who was direct and to the point and composed his nasty responses to Newton in just a few hours. Newton however, being exacting, scrupulous and vigilant in all that he did, spent months carefully scribing his well thought out replies. It is said that he once spent months bitterly torn up by Hooke's comments. Mulling them over and over in his mind, whilst he viciously responded Hooke's cutting correspondence. This Caused him far more internal damage than his words could ever hurt Hooke. In later years they fought again over planetary motion and Hooke's lashed out with an accusation that Newton had stolen some of his research and included it in the Principia. Newton was furious.

In 1684, Newton started a brutal verbal altercation with a German mathematician called Gottfried Leibniz. Originally instigated by the publication of a mathematical paper by Leibniz, in which he tried to solve the 'mystery of nature', incorporating mathematical expressions into his thesis. Newton reacted with a statement that he had already done the same work almost 20 years before and the German philosopher had copied and stolen his work.

During the years following the publication of the Principia, Newton had become disenchanted with Cambridge and all the petty arguments and eternal disputes he had become embroiled in. He decided to move to London to take up a role he had been offered at the Royal Mint in 1696. This turned out to be a permanent relocation to the capital, however antagonism followed his path.

Further conflict erupted with Leibniz when he claimed that he had discovered infinitesimal calculus first. Leibniz insisted that Newton had copied his ideas. In 1712 The Royal Society eventually raised an investigation into the matter. As Newton had been the president of the society since 1703, Newton got his way and the society backed him in full and declared Newton as the man who had defined Calculus. In later years it was determined that the two mathematicians had probably made their discoveries independent of each other.

These bitter encounters with his fellow peers affected Newton quite severely and at one point drove him to retreat from the scientific community almost completely. Newton's introduction to the social side of science, public debate and discussion of ideas was a disaster. He simply couldn't deal with the dents to his ego and the resulting emotional trauma. His frustrations and unhappiness, on top of the death of his mother, caused Newton to shy away. He was now bored with mathematics anyway and by the mid-1670s, still only in his early 30s, but with long locks of unkept grey hair, he placed himself in a state of total isolation. As usual Newton had his own agenda and although during this 6 year period he published very little, he turned his focus and his 100-hour working weeks towards two of his other interests. Subjects that he was keen to keep to himself. He really didn't want to discuss either of them with anyone. So he secretly set about the mathematical and textual analysis of the Bible and the mysteries of alchemy.

During the earlier part of his life, Newton was really rather introverted and extremely protective of his privacy. Even as he matured, having become famous, rich and bestowed with honours and international acclaim as one of the world's foremost thinkers, he remained deeply insecure. He suffered from depression, a violent temper and could be incredibly vindictive when he felt that his ideas were challenged or threatened in any way.

It is thought that Newtons early experiences such as the loss and betrayal of his mother, permanently damaged Newton's ability to trust and get close to people. It has also been suggested, although unconfirmed and often disputed, that Isaac was a repressed homosexual. Which if true would have caused a man of his background and upbringing to unduly experience extreme mental strain. Presenting conflict in his mind between his natural urges, his religious views and pious mindset. Had he been born in the 21st century, this dilemma and frustration would never have existed.

Rumours of his underlying homosexuality arose due to his contact with a brilliant Swiss mathematician named Fatio de Duillier. It seems Newton was very fond of young Fatio who he met in 1690 at the age of 48. Somehow Fatio managed to bypass Newton's defensive abruptness and reach the person beneath. Much has been written regarding their relationship and sudden break in 1693.

There are written records of meetings between the two men which show that Isaac first went to visit Fatio in London in September 1691. In the Autumn of 1692 Fatio visited Newton in Cambridge and for the duration of January and February of 1693, Fatio stayed with Newton at his home in Cambridge. They may have met many more times not recorded, but the final note about their time together shows that Newton stayed at Fatio's in London for the whole of May and June of 1693. The following July Newton suffered an emotional breakdown. We can't be sure if this was the result of an altercation and break up with Fatio, or some other reason, but he never saw Fatio again.

Newtons closet colleges, John Locke and Samuel Pepys revealed the depth of Newton's sorrow and depression in July 2693, by describing it as his 'discomposure in head or mind or both'. Newtons breakdown was greatly compounded by his psychological paranoia which induced a mistrust of people. He said after five nights of sleeping 'not a wink', he temporarily lost his mind and convinced himself that his friends Locke and Pepys were conspiring against him. However, he managed to regain himself and by the end of the year he appeared to have made a full recovery.

Prior to Newtons entry to Cambridge, he kept a notebook known as the Pierpont Morgan Notebook. In this book he wrote 10,995 words about various subjects. At one point in the book there are groups of words set out in lists, under a number of subject headings. Next to some of the words Newton makes some interesting and revealing notes. The word 'Father' is followed by 'Fornicator and Flatterer'. Next to 'Brother' are the words 'Bastard, Blasphemer, Brawler, Babler, Babylonian, Bishop' and the final word is 'Benjamite'. According to the Bible, Genesis 42, Jacob gave his youngest son Benjamite, special attention and favouritism over his other sons. The point being that Isaacs half-brother was called Benjamin.

In another notebook dated 1662, Newton wrote a list of all the sins he could remember having committed to date. It's a personal insight into how religion ruled his early years. Being addressed directly to God, he lays his wrong doings out and confesses his most awful behaviour. What is quite endearing is how short the list is and how trivial many of his sins are. Yet they consumed his mind with guilt years after he had committed them. Written in shorthand code, as only God could know his sins, they reveal a fascinating glimpse into Newton's conscience.

His sins included minor instances of sabbath-breaking, including 'squirting water on thy day'. Other misdeeds recorded are 'Idle discourse on thy day', by which he means the Sabbath which is Sunday. According to the Christian faith the Sabbath is a day of prayer and reflection. It is a sin to do any kind of work or have trivial thoughts on the Sabbath. He goes on 'and at other times, Peevishness at Master Clarks for a piece of bread and butter'. Other behaviour he described is slightly more serious and reveals traits that would develop in his adulthood. It shows glimpses of his deep set temper and the black depression that he would later suffer from. He confessed to 'Striking many, punching my sister, wishing death and hoping it to some'. It says much about the sternly puritanical cast of Newton's upbringing and the unforgiving nature of his beliefs.

The pain and sense of insecurity that his mother's desertion of him as a child, rendered him obsessively anxious during times when his work was published. He often became irrationally aggressive and violent when he defended it. This all righteous attitude and assertive behaviour accompanied Newton throughout his entire life. He was at the mercy of others as a child. Not in control of his circumstances, vulnerable and dictated to. As an adult he was determined to be in control of everything. If his Father had lived and his Mother remained with him, Isaac might have been a happier human being. However, he might not have pursued all his scientific quests with the vigour and determination with which he did and the world would have missed out on the wealth of amazing discoveries he provided.

Isaac Newton was suffering long before he knew the word 'suffering'. His childhood needs were basically met but warmth, love, appreciation and a feeling of belonging were all lacking. The little boy Isaac was isolated and lonely and more than anything, rejected. He survived by closing down his emotional expectations in life. By growing a thick skin with a prickly, hard exterior. Cutting off his own emotions and any feelings of compassion for others. This became more evident as his power grew and he displayed a lack of sympathy for anybody. Described as Cold and calculating, cunning and quick-tempered, he was not the most charming man.

Newton never married. It appears he failed to form any warm, long lasting friendships at all. Although in 1689 he met the noted Philosopher John Locke, who is recorded as being Newton's closest adult connection, aside from Fatio. How close their friendship really was, is unknown. Although Newton was said to be generous with his time and money, he was emotionally baron and found it difficult to give any kind of emotive response to family, peers or acquaintances. He existed within a narrow strip of life, dedicated only to his vital work. Rarely venturing outside Cambridge, London, or his home at Woolsthorpe. He lacked humour and frivolity and was never able to laugh at himself. The only emotion he seemed able to muster was anger, which made him unbalanced, spiteful and vengeful.

Though his personality failed to endear him to virtually anyone, it served his career remarkably well. Ruthlessness is an unexpected trait for scientific mastery, but whenever challenged Newton pursued personal vendettas against his enemies with unrelenting vigour. Grudges against both Hooke and Leibniz persisted even after their deaths. Newton continued to tarnish the reputations and discoveries of both men, whilst promoting his own and ensuring his place in the history books. Newton once said "Tact is the art of making a point without making an enemy". Clearly he knew what tact involved, however it seems he was unable to actually be tactful, ever.

Strangely, Newton was made Master of the Queens Royal Mint in 1700 and during this time the truth of his childhood damage became even more apparent. He set about clamping down on those that forged the Royal coins. At that time almost 10% of Britain's currency was found to be forged. Newton was savage and extremely successful in this role, which is hardly surprising when you consider the way he punished those he caught in the act.

His obsessive nature made Newton a relentless pursuer of wrong doers. It is said that the famed inventor took to the dark, mean streets of London in full disguise, to root out counterfeiters and bring them to justice. He took his role as Master of Mint very seriously and prosecuted and punished the forgers he had caught under his new strict guidelines, in the most severe way possible. In those days a currency coin was actually made out of gold and one way of creating more money was to nibble bits of gold from existing coins and then remould the bits into a new coin. People who carried out this particular method of theft were called 'Coiners'. One notorious and difficult to capture coiner called William Chaloner, under Newton's instruction, was hanged at the gallows and publicly disembowelled. Newton would go on to hang or burn at the stake a further twenty-eight counterfeiters and was quite famous for coldly refusing pleas for clemency.

It's unclear how Newton managed to balance these violent and murderous acts with his pious Christian beliefs. The 6th commandment in the Holy Bible clearly states, "Thou shall not kill". Maybe he justified it with the idea that he did not actually carry out the killing himself, leaving that unwholesome, nasty job to the hangman and prison staff. However, it was he that dictated the penalties and from records we can see that he thoroughly enjoyed his position of power at the Mint, where he could indulge the darker side of his nature. Somehow, he killed many people, without damaging his scrupulous puritan conscience. The 8th commandment says "Thou shall not steal", clearly a lighter offence in the eyes of God and yet Newton, who constantly referred to Gods presence, disregarded one of the only ten rules God gave.

Picture Newton standing in some elevated position, with a birds eye view of the scaffold, as the Mater of Mint and dictator of punishment would have. He looks down on a public square where hundreds of onlookers have gathered and captured coiner, William Chaloner stands with a noose around his neck. Newton observes as William is hung by the neck, he squirms and chokes and after many minutes is about to die. Just before the relief of death, he is cut down, laid on a bench and whist still recovering from strangulation, Newton watches as Williams abdomen is sliced open. He is still alive and his bowels are gruesomely dragged out of his body and spill all over the floor. He screams with agony until death arrives and then Williams body is cut into quarters and stuck on spikes around the town as a warning to the masses. Newton walks calmly home, sits down, picks up his Bible and pats himself on the back for a job well done. His hypocrisy is staggering when you remember that William didn't kill anyone.

He was only guilty of fiddling a small amount of gold to feed his family. Newton was guilty of cold blooded murder and the arrogance to take on God's role, of deciding who lives and who dies.

Further pharisaic behaviour is evident when considering Newtons ability to justify his alchemists pursuit of the Philosophers Stone, involving the attempted production of gold. An act which was just as illegal as coining at that time. Newton really did view religion and the law as having one set of rules for the masses and another for the divine few, such as he.

Aside from childhood neglect, another theory relating to Newtons challenging or detached behaviour which has now gained followers, is that Newton might of suffered from undiagnosed Asperger's Syndrome. The four main indicators of Asperger's syndrome are -
- Social impairment shown by poor non-verbal
  communication.
- Inability to establish friendships and lack of empathy.
- Little regard for communication with society.

- Focused dedication and compliance with routine.

It is clear that Newton suffered from all of these symptoms during his life, so Asperger's might explain some of the problems he experienced relating to other people.

Knighthood & Honours

Long before his revolutionary work Philosophiae Naturalis Principia Mathematica was published, Newton was already considered one of England's leading thinkers. In 1669 he had been awarded the highly respected position as the Lucasian professor of mathematics at Cambridge, following in the footsteps of his great mentor Isaac Barrow. Later geniuses to hold this position include Charles Babbage, also known as 'The Father of Computing', Paul Dirac and more recently Stephen Hawking in 1979, who studied the basic laws governing the universe. It is now considered the most famous academic chair in the world.

Newton's position as Master of Mint was quite an endeavour and a great achievement. He didn't just walk into the job. He had exhibited interest in the position which was based in

London, but his application was not immediately as successful as he had expected. He was reduced to taking up the more minor position as the Warden of the Royal Mint in 1696. He held this title for three years until he became Master of Mint in 1699. As leader of the Royal Mint, Newton exercised his power and in 1717 he played a significant role in recovering Britain from a financial crises. Newton set about recalling all the old currency and issued a more reliable one, by actually changing the English pound from the sterling silver standard to the gold standard. He held his post at the Mint until the end of his life.

Newton also served as a member of Parliament in England from 1689 to 1690 and again from 1701 to 1702. However, regardless of the fact Newton was a genius, he didn't turn out to be much of a politician. During his two separate one year stints as a Member of Parliament, it is said he spoke up only once and that was to tell someone to close a window. It seems strange that such a strong minded and opinionated man could think of nothing to say about the running of the country, for two whole years. Even more strange, is the fact he became an MP in the first place.

Further honours were bestowed upon Newton when he was elected President of the Royal Society in 1703. In 1705 he was knighted by Queen Anne and thereafter was known as Sir Isaac Newton. He was only the second scientist to be knighted, after Sir Francis Bacon. Bizarrely Isaac Newton did not become Sir Isaac Newton for his many magnificent contributions to the field of science and all his vast scientific achievements. The knighting of the famous English physicist and mathematician was to better his chances as a politician, by boosting his political profile. On a visit to Trinity College in Cambridge, Queen Anne knighted Newton in the university's Master's Lodge on April 16th 1705. At this time Newton was 63 years of age and he had been elected as one of the two members of Cambridge University's parliament twice before. He was now competing for the seat for a third time. Newton had been advised to seek assistance from Queen Anne as such an endorsement would assist Newton to gain public favour and win the seat again. As it turned out, Newton's knighthood didn't swing the public and he actually finished last out of the four candidates standing. Bringing to an end Newton's pointless political pursuits.

*"Gravity explains the
motions of the planets, but
it cannot explain who sets
the planets in motion".*

Isaac Newton

Chapter 5

Science Vs Religion

Alchemy played a massive role in Newtons life and in spite of all his scientific work, it is said that he devoted over half his adult life to the subject. As well as various, sometimes odd, religious endeavours. As was his way, these were not mere ponderous frivolities, but vast, in-depth, full investigations. Although he had beliefs in conventional religion, he joined a heretical sect called Arianism which held specific beliefs that the Holy Trinity did not exist. A belief shared by Mormons and Jehovah's Witnesses. Although ironically, he didn't seem to mind studying at Trinity College, Cambridge.

Newton was open minded about the possibilities of various beliefs. He considered many religions even learning Hebrew to help him in his endeavour to better understand original texts relating to the lost Temple of King Solomon, which had stood in Jerusalem. He studied floor plans and other related documents on a mission to prove they held mathematical clues, as to the date of the end of the world and the second coming of Christ.

The Christian faith and the desire to go to Trinity College initially entered Newtons life via his maternal uncle William Ayscough. Who, as well as being Isaac's school headmaster at Grantham, was a Trinity College graduate and also the rector at Burton Coggles, a small village 5 miles East of Newton's home in Woolsthorpe.

Another religious influencer came in 1661 when his Mothers best friend's brother, Mr Humphrey Babington. He was a recently reinstituted fellow at Trinity and was made rector of Boothby Pagnell, another little village just over six miles from Newton's home. Although there are few known confirmed facts about their relationship, it is known that Babington was close to Newton in his early years at the college and he almost certainly acted as a patron for Isaac during this period.

Even though later in life, Newton was at odds with some aspects of the Bible, many of his quotes indicate that his faith in Christianity played a central role in his life stating –

"This most beautiful system of the sun, planets and comets,

could only proceed from the counsel and dominion of an intelligent and powerful Being".

This reference could be applied to God or aliens, but it is clear that Newton believed that there was a creator of all that he beheld. Of those that didn't believe in anything he said –

"Atheism is so senseless. When I look at the solar system, I see the earth at the right distance from the sun to receive the proper amounts of heat and light. This did not happen by chance".

Although there is obviously some overlap between them, it is clear that there is a distinction between Newton's main core religious beliefs and his technical theological research, relating to the study of the nature of God. Both seen by Newton with mathematical equations behind them. The latter consisted of his study of prophecy, the nature of God and the nature and historical role of Jesus Christ. As well as the form and function of pre-Christian religion, the evolution of the Christian doctrine and the documented history of the Bible.

He wrote up much of this work in the form of treatises, a written work dealing formally and systematically with the subjects. Many of which are as original and monumental as the works in the exact sciences for which he is best known. The fact that they are not part of the major religious writings of the day is a direct consequence of the views that he had expressed. Since his views were radically heterodox, not conforming with accepted or orthodox standards and beliefs of the day. He would have been condemned and considered heretical by the Church of England. Thus Newton decided to suppress and hide these documented views from public scrutiny, for the sake of his own protection.

His writings show he was under the illusion that he had a direct connection with God and that God had chosen him specifically to interpret the Bible. Having done this, on God's behalf, he concluded that the world would end no sooner than 2060. He declared -

"The time times and half time do not end before 2060. It may end later, but I see no reason for its ending sooner. This I mention not to assert when the time of the end shall be, but to put a stop to the rash conjectures of fanciful men who are frequently predicting the time of the end and by doing so bring the sacred prophesies into discredit as often as their predictions fail. Christ comes as a thief in the night and it is not for us to know the times and seasons which God hath put into his own breast".   Isaac Newton

His prediction has stood solidly for over 300 years and the date 2060 is precisely 1,260 years, after the foundation of the Holy Roman Empire. This prediction was discovered quite recently in paperwork that had been rejected by the Royal Society. Newton also made a specific prophecy about the second coming of Christ, which he said would occur just four decades from now.

In the first twenty years of his life Newton was exposed to opposing religious traditions. The friendly influence of three clergymen, his uncle William Ayscough of Burton Goggles, William Walker of Colsterworth and Humphrey Babington of Boothby Pagnall, all affected Newton. However, the influences bearing upon Newton were not exclusively Puritan, as two of the three clergymen mentioned above were stout Royalists and Anglicans. William Walker, was one of the most powerful parliamentarian figures in Grantham. Although, the senior and most influential male figures in his young life were ordained members of the Church of England.

At some point in the 1670s Newton came to the view that the simple and authentic form of Christianity had been corrupted in the centuries following the life of Jesus Christ. Producing a brand of religion that was now accepted as orthodox by the Roman Catholic Church and to some extent, by the Church of England. He concluded that the orthodox notion of the Trinity was a fiction that had been invented in the early fourth century and subsequently promoted by servants of the devil. The Christian doctrine of the Trinity holds that God is one God, made up from three coeternal persons, that exist with each other eternally. They are consubstantial, meaning they are of the same substance or essence and hypostases. Each of the three persons of the Trinity, specifically the Father, the Son (Jesus) and the Holy Spirit, are as 'one God in Three Divine Persons'. The three persons are distinct, yet are one. Newton rejected the Holy Spirit, he believed that it was a wrong to give to any other being the worship that was properly due to God and that

Jesus was divine but was not God. Therefore sections of the Bible distressed Newton, as the bible makes many references to the Holy Spirit, for example –

Genesis 1:2-3, "When the earth was created but still in darkness and without form, the Spirit of God was hovering over the waters". The Holy Spirit is the 'breath of life' in creation: "Then the Lord God formed a man from the dust of the ground and breathed into his nostrils the breath of life and the man became a living being."

You can see why Newton as a scientist, had difficulty with the Holy Spirit and in fact with religion altogether. On the one hand he believed in God the creator and yet his urge to find a logical and scientific explanation for all, was instinctive. On some days logic and science won and on others days God won Newtons favour. He declared "all variety of created objects which represent order and life in the Universe could happen only by the wilful reasoning of its original Creator, whom I call the Lord God."

It isn't possible to say exactly when Newton started looking into the radical investigation that would lead to his dissident position on the Trinity. There is no evidence that he was raised in any radical way. Newtons early exposure to both the traditional Church of England and Presbyterian influences when he was a teenager, doesn't appear to be the source of his strictly puritan moral attitudes. The evidence indicates that his heresy was a result of his extreme Protestant dislike of Roman Catholicism, which was revealed when he objected strongly against the attempt of James II to admit Catholics into Cambridge University.

As he viewed it, the world was contaminated and only a small selection of pure Christians such as himself, were of the true faith. The possibility of saintliness lay only with the divine few. Newton firmly believed that the Bible says that only pious men would be shown the truth, if they really looked for it. If they specifically sought out the clues, that would be missed by merely reading the book. Somewhere in the detail lay a truth that Newton sought. He was certain that pious, great men throughout history had passed down these secrets of the Universe by coding and hiding them in their scripts. Newton was convinced that one alchemist in particular, the Swiss physician Paracelsus, was one such man.

He also believed that he had been selected as a special person that had been drawn to the work of Paracelsus, in order that he could recognise and translate all the divine information locked inside. But it was information that he and he alone could be privy to and it was of great importance that he protect and conceal it from the irreverent masses. It is said that after concluding his own Law of Gravity he claimed that Plato, Moses and Pythagoras all knew about it first. Although he failed to mention that in his book the Principia.

Regardless of how special Newton thought he was with his ever divine presence, evidence from some early eighteenth century writings suggest that Newton was content to remain in the Church of England. He is known to have worshipped regularly at his university church, Great St. Mary's, in the mid to late 1680s. However, he was secretly waiting for it to become so enlightened and properly Christian, that someone with views such as his own, could truly worship within its fold.

If he had published or even publicly voiced these ideas at that time in the late 1600's, he would have been forced to leave Cambridge University. It would have been impossible for him take up the senior political and administrative roles he was awarded in the 1700's. He might not have written the Principia, his most famous and influential offering to the world. He wouldn't have risen up to become the respected cosmopolitan, scientific public figure we know about today. He would never have been promoted to Master of the Mint in 1699, nor would he have been made President of the Royal Society in 1703 and he definitely wouldn't have been knighted by the Queen, two years later.

Newton was wise in that he patiently guarded many of his real thoughts on religion, whilst waiting for the world to catch up with his beliefs.

At times of depression Newton hallucinated and had conversations with absent people.

"Plato is my friend,
Aristotle is my friend,
but my greatest friend is
truth".

Isaac Newton

Chapter 6

Newton & Alchemy

Today when people think of alchemy, they generally think of it as some sort of witchcraft or sorcery. Few would accept it as true scientific endeavour. However, in Newtons day it was practiced by many famous scientists and it actually laid down the foundations of basic Chemistry.

The word alchemy originates from the Arabic *al-kīmiyā*, which translates as 'the art of transmuting metals'. The true and actual definition of alchemy is the process of taking something ordinary and transforming it into something extraordinary. Sometimes in a way that cannot be explained. One of the main concepts of alchemy was founded on the idea of transmuting lead into gold. Transmutation is the action of altering or transforming into another form. Most often this referred to changing base metals into noble metals or vice versa. However this wasn't always the case, as transmutations can come in many forms. This strange work was once thought to be a spiritual transmutation, since it involves changing the inner state of a substance.

With alchemy spanning across so many different countries and cultures throughout time, it's hard not to wonder if there is something in it. Although, where does one even start to turn lead, which we now know has atomic number 82, into gold with atomic number 79, which are defined as elements by the exact number of protons they contain. Changing the element, requires changing the atomic (proton) number.

From antiquity, we have known people to master apparently magical transformations, for example by heating sand to high temperatures to create glass. You can understand why, upon seeing that kind of phenomenon at that period in time, people believed that it would be possible to change other substances into gold.

To Newton, alchemy presented a conundrum, imbued in ancient Greek myths and the work of the great alchemists from the past. Newton believed that, if he could solve the conundrum, he would be able to understand and control nature. His belief that he alone could control nature came from his arrogant perception that he was divine and one in a sequence of special men specifically selected, to receive this sacred, secret, ancient wisdom. Newton was so committed to alchemy and its mystical endeavors that he gave himself the name Jehovah Sanctus Unus. Which when translated, Jehovah means 'The Holy One'. Newton's ego and pure belief in himself was really quite astonishing and it increased with every success and award he received.

The first rule of alchemy states 'You cannot obtain anything without first giving something in return' and was based on the belief that there are four basic elements in nature: Air, Fire, Water and Earth. It was founded on a complex spiritual worldview in which everything around us contains a sort of universal spirit. Metals were believed not only to be alive but also to grow inside the Earth.

Alchemy first developed as a subject in Alexandria. Egyptian metal crafters and fabric dyers wanted to understand more about their products around the start of the Christian faith. True alchemy did not flourish until philosophy added further dimensions to the subject. Plato and Aristotle added philosophy to earlier interest in the elements. Both Aristotle and Plato had their interest first ignited by Empedocles, a Greek philosopher and a native citizen of Akragas, a Greek city in Sicily. He was the man who had first established that the Universe was made up of four fundamental elements, Earth, Water, Air and Fire. Plato refers to Empedocles explicitly when he claims that all natural objects are made up from those four entities. Empedocles provided the origins of the concept of elements which passed on to Plato, then to his pupil Aristotle. In turn Newton had also been introduced to their work, through his alchemical studies and his exposure to classical research at Cambridge.

Newtons fascination with alchemy devoured vast chunks of his life. His time as a boy living with Mr Clark the apothecary (chemist) had not been wasted. He wrote more than one million words about the subject throughout his time on this planet. Mainly in the hope of using ancient knowledge to better explain the nature of matter, everlasting life and possibly endow the man with vast riches.

Prior to the formation of one standard, uniformed language

for chemistry, scientists or alchemists used symbols and signs rooted in ancient astrology and mythology to record and portray their practices, methods and the substances they used. As a result, even a basic experiment read like a spell.

The first step of every alchemy formula/ spell/ experiment is to LIGHT your cauldron.

The second step is to gather the required ingredients, which could be any combination of alchemic elements.

The last step of every formula is to ALCHEMY SEAL.

Ancient alchemists did not think about elements the way modern Chemists do. They used the word 'element' to refer to anything that was basic. In the beginning ancient Greek philosophers and alchemists thought that everything on the planet consisted of only the four elements. As the centuries passed understanding developed and the number of elements slowly grew.

The following list is just a sample of alchemy elements and the symbols, commonly in use since the late 1600's. You can see as you read on that it is the Periodic Table in the making.

Earth was considered a combination of water and air and was symbolised in alchemy by a downward triangle, divided by a horizontal line crossing through it. Dryness and cold are properties that Plato connected to the symbol for earth. The alchemical earth is also considered as a symbol for physical sensations, movements and was represented by the colours green and brown.

Fire was symbolised by an upright triangle and was one of the basic elements found in alchemy. The alchemical fire was associated with warmth, heat and dryness by Plato and it was often associated with the colours red and orange. It was said to symbolise emotions like love, passion, compassion, hate and anger. Any emotion we would consider 'fiery', or with a 'rising energy', which has been associated with aspiring to reach the divine above us. The element of fire was considered masculine.

Water was given a downward triangle to represent it's use and it was the opposite of the fire symbol. Again, as a reverse of fire, water was considered feminine. The downward triangle has been used by many cultures as a symbol to represent women and femininity. It was also associated with mercury and thought to correlate with intuition. The Greek philosopher Plato linked water with its characteristics, such as cold, moist, wet and the colour blue was linked to this element.

Air is the fourth classical element amongst the symbols. An upright triangle with a line across it was defined as the sign for this element, which is the exact reverse of the alchemists earth symbol. Plato associated wind or air with properties like warmth, wetness and he gave air the colours blue, white and grey. The air symbol in alchemy was considered to represent the Holy Spirit and life-giving forces, such as the sources of life, like breaths.

The philosopher's stone was an age-old substance in alchemy, which many believed was the thing that would actually turn common base metals like lead and mercury, into rare and expensive ones such as gold and silver. In addition many believed that it was the elixir of life, having rejuvenating powers that could be used to make one immortal. Creating the stone was a major and prized goal in alchemy and Isaac Newton made great efforts to discover the enigmatic formula. The Philosopher's Stone represented enlightenment, heavenly bliss and perfection in the world of alchemy. Uniquely capable of the transformation of physical substances from a state of imperfect, temporal existence into one of spiritual perfection. It's no wonder that for centuries, the Philosopher's Stone obsessed so many alchemists. The symbol was made up from an outer circle, with an inner triangle, then a square and a smaller circle at the centre. According to rediscovered manuscripts handwritten by Newton the recipe for the Philosophers Stone involves –

"combining one part Fiery Dragon, some Doves of Diana and at least seven Eagles of mercury".   Isaac Newton

He believed this would create the key precursor to the Philosopher's stone. The manuscripts have been held in a private collection for decades but now the 17th-century documents are safely in the hands of the Chemical Heritage Foundation. Which is a non-profit organisation based in Philadelphia, Pennsylvania. The dedicated group bought the manuscripts and are currently working on uploading digital images and transcriptions to an online database, so more people can study Newton's beliefs and views on alchemy.

From all evidence found, it appears that Newton failed to successfully develop the Philosophers Stone, but his quest was not a complete waste of time. Some real insights we still use today, emerged from his proto-chemistry experiments. Alchemists invented distillation, proto morphine (laudanum), oil paints, inks and much more.

 Alchemy involved a group of elements called the **Three Primes or Principles** (Tria-Prima). Which included **sulfur** (sulphur) relating to the human mind, **mercury** associated with the spirit and **salt** which refers to the body or base matter. Mercury, salt and sulfur each had their own individual symbols, but the Three Primes symbol represented all three elements as a group.

 Sulphur (Sulfur) is one of the most interesting elements in alchemy, also known as Brimstone. It is represented most commonly by a symbol with a triangle standing on top of a Greek cross, with arms of equal length, which predates Christianity. This symbol came to be known as Satan's Cross or the Leviathan Cross, after it was adopted as a Satanic symbol by Anton LaVey in 1960s. This symbol is still used today by people of Satanic faith.

Sulphur was associated with properties like heat, dryness and masculinity. It was thought to represent dissolution, evaporation and expansion or expansive forces in alchemy. Solutions with sulphur were used to speed up the healing of wounds. Ancient Greek alchemists used it in medicine to treat haemorrhoids, arthritis, hypothyroidism (dry skin and brittle nails) and in homeopathy, by using minute doses of this naturally occurring substance.

Sulphur has very distinctive physical, as well as chemical properties. It often occurs as a brilliant yellow powder. When burnt it produces a clear blue flame and a strong distinctive odour. Sulphur can sometimes be seen as a bright yellow layers on the top of the earth or it can be mined as an ore. An ore is a naturally occurring mineral used as a source for an element.

The Bible is full of references to brimstone. Sodom and Gomorrah were two towns which God is said to have destroyed because of the sinful behaviour of their people.

Genesis 19:24."The Lord rained upon Sodom and upon Gomorrah brimstone and fire".

Today, the world makes more sulphuric acid than any other chemical and it has a vast number of important, modern day uses. Sulphur is involved in the production of various drugs, explosives, dyes, detergents and fertilisers.

☿ **Mercury** is the second part of the Three Primes. This element was represented by an upward pointing crescent moon on top of a circle on a cross. Mercury was considered a symbol of femininity and we still use this sign to indicate females in our modern society. Also known as hydrargyrum or quicksilver, mercury was said to represent the spirit of life or the life force. According to alchemy, this spirit was not lost when matter shifted between a solid and a liquid state and it even transcended death and continued to exist beyond it. Mercury was regarded as moist, wet and cold.

Mostly mined in Spain from the Almaden mine, mercury can also be found in Yugoslavia, the USA and Italy. It is obtained from an ore called cinnabar and this element was named after the Roman God Mercury. It is one of the few element to retain its original alchemical name in modern science. Vials of pure mercury have been recovered from Egyptian tombs dating from 1500 BC, but was in use by ancient civilisations as far back as 2000 BC.

Salt is the third and final prime in alchemy. It is represented using a symbol comprised of a circle divided in two by a horizontal line crossing through its middle. Salt represents condensation and crystallisation. It was also considered as a symbol for physical matter and the basic essence of all things in nature. Since Paracelsus 1493-1541, salt has played a role in alchemy. Paracelsus was one of the first medical men to believe that physicians should have a good knowledge and extended understanding of the natural sciences, especially chemistry. He was a pioneer of medicine and lead the way by initiating the incorporation of chemicals and minerals in medicine.

Salt comes from weathering rocks and volcanic activity. When the seas first formed at the beginning of Earth's existence, the vast waters washed against the rocks and a weathering process began, which leached or dissolved the soluble elements out of the rocks. Humans have been finding uses for salt since the beginning of time, before recorded history began. The earliest mention of the harvesting of salt was in China and in Eastern Europe over 8000 years ago. People initially boiled water from lakes and springs to extract salt for its healing properties.

Later they moved onto the sea. The process for extracting salt from the sea is an ancient method involving evaporation ponds. Small, shallow pools were dug out and joined by a narrow canal to the sea. Using a wide area of shallow water allowed the water to absorb more sunlight and evaporate quicker.

At one time salt was extremely valuable. The word 'salary' comes from the Latin word 'salarium', which translates as 'salt money'. The Romans paid their soldiers an allowance of salt and as a result the word 'salarium' was the word used for military pay, long after salt was used to pay soldiers. The Chinese also valued salt highly, so highly that in 2200 BC, Emperor Hsia Yu levied the first tax ever and it was on salt. In ancient Greece, slaves were often sold for salt and a poorly behaved slave was 'not worth his salt'.

The Egyptians used salt to preserve and season their food. Salt extended the edible life span of fish and other meats, making it possible for the Egyptians to accumulate surplus food and improve the country's economy through domestic and foreign trade. They also used vast amounts of salt during the vital, drying out process of mummification.

Silver is one of the seven metals in alchemy. It was most often represented by the crescent moon. Although the moon was widely associated with silver in alchemy, the crescent moon symbol was also used to represent the actual Moon. Native Silver is mined from the Earth's crust in a pure, free elemental form and is a white, soft, shiny transition metal. It possesses high levels of electrical conductivity as well as thermal conductivity. In most cultures silver has been shaped and polished to create spectacular jewellery and it was also used to produce coins for currency in many countries.

Gold was associated by alchemists with the sun and a state of perfection. It was defined by a circle within a circle and represented spiritual, physical and also mental perfection for human beings. Gold was always one of the most prominent alchemical symbols. Alchemists believed that elementary gold represented a pure consciousness and was used to purify the philosopher's stone. It has been described by alchemists as the divine, creative influence present in all matter.

Many ancient cultures used gold to make jewellery and art because of its impressive aesthetic qualities, ductility and malleability. Evidence the ancient Egyptians adored it go back to 3,000 BC. Gold once played an important role in ancient Egyptian mythology and was valued highly by pharaohs and temple priests. Gold has also been known as a sign of immortality among the gods and of wealth among ordinary people. Since gold can be found all over the world, it has been mentioned numerous times throughout ancient historical texts.

Lead is another one of the seven metals used in alchemy. Lead is sometimes represented by more than one symbol and has been associated with the planet Saturn. It is the heaviest of the seven metals and is tied to gravity, form and manifested reality. It is an extremely resilient metal and is known for its durability and resistance to change. Lead products dating from 7000 BC are still intact and lead water pipes installed by the Romans 1,500 years ago, are still in daily use in some places today. Alchemists sometimes depicted lead in their drawings as the God Saturn or a crippled old man with a sickle, Father Time or a skeleton representing death itself. Any of these symbols, found in their manuscripts meant that the alchemist was working with the metal lead in their laboratory.

Although alchemists considered lead as the lowest of the base metals, they treated it with a great deal of respect. Lead was said to carry all the energy of its own transformation and it was that hidden energy that the alchemists sought to free. To the alchemist, this ancient metal was considered to be a powerful 'sleeping giant', with a dark and secret nature that encompassed both the beginning and the end.

Lead was one of the first metals in history to be mined due to its ease of extraction from naturally occurring ores. It was prevalent from as far back as 8,000 years ago in most of the ancient civilizations. Being abundant, resistant to corrosion, soft, easy to shape and durable over extended periods, it was a favourite during the Iron Age. Alchemists found it easy to use due to its relatively low melting point and spent most of their time trying to turn it into gold.

♀ **Copper** was linked by alchemists to the planet Venus. The symbol for Venus was deemed to be a female symbol. The same sign was used for the Goddess and also for the planetary symbol of Venus. As such, copper was said to embody the characteristics of love, balance, feminine beauty and artistic creativity. Finding small amounts of copper was relatively easy in its metallic state in many countries of the ancient world. Copper was the first metal used, more than 10,000 years ago by ancient man. It has been a vital material for human-kind since pre-historic times. One of the major "ages" of human history is the Bronze Age and bronze is a copper alloy (a combination of two or more metallic elements).

The word 'copper' is said to have come from the Latin word 'cuprum', which translates as 'from the island of Cyprus'. However, Cyprus is not where most ancient evidence for the use of copper has been found. The shiny orange, red or brown metal was used in the Balkans and the Middle East from about 8000 to 3000 BC. Egypt and Europe followed later and began to make their own copper implements.

Copper metal has great transfer and connecting powers as it easily combines with other metals and it readily transfers warmth and in our time, electricity. Alchemists used Copper in rituals, spells and amulets to try to promote sensuality, friendship, positive relationships of any kind, negotiations and peace.

Antimony was represented by an upturned version of the female symbol. It was considered to be a symbol for the wild and free parts of human nature. Antimony compounds are silvery grey and have been used since ancient times in a powder form in medicines and makeup. Ancient Egyptians used it to define their eyes in a very distinctive way. Better known today by the Arabic name Kohl, it has many other modern applications.

Antimony is sometimes found free in nature, but is usually obtained from ore. Its use has been dated back to 2500 BC, via an anatomy-plated object. Although it is poisonous and causes symptoms similar to arsenic poisoning, it is far less toxic than most of the heavy metals and was well known to medieval alchemists for use in their potions and spells.

Tin had connections with the planet Jupiter. It acts as a catalyst when oxygen is included in a solution and helps to accelerate chemical attack. Tin was associated with exhaled breaths and viewed by alchemists as a life force. The symbol for tin includes a straight horizontal line, connected to a curve line and crossed by a vertical line. The meaning was thought to be that each part, placed together is more powerful than each part stood alone.

Tin was known to many ancient cultures and is mentioned in the Bibles Old Testament. The metal was too soft for most practical uses but early metal workers mixed it with copper to make the alloy bronze. Then its acquisition became an important part of ancient cultures from the Bronze Age onwards and it was used widely in the Middle East and the Balkans around 3000 BC.

Iron was said to represent the planet Mars. The symbol for iron was a circle with an arrow pointing out to the right. It resembles the symbol for masculinity and hence connotes physical strength and male energy.

Iron is the most common element found on planet Earth, comprising 5.6% of the Earth's crust and was well known to the ancient world. The process of working iron began in Europe in 11 BC and its use slowly travelled to other continents over the next century. The discovery of iron came long before the Iron Age but it gradually started to replace bronze in the production of weapons and tools. Initially iron was used for ornaments in the Middle East around 2500 BC. It started to be used on a large scale by the Hittites, in an area we now know as Turkey and Syria about 1200 BC. Beads made from meteoric iron were found in Giza in Egypt dating from 3500 BC.

Objects made of iron from ancient times are far less common than objects crafted from silver or gold because of the speed with which iron corrodes. Methods for working iron did not come fast and even after the development of smelting, centuries had passed before iron replaced bronze. When iron eventually found it's place, it became the primary metal in the production of tools and weapons. The Iron Age commenced between 1200 B.C. and 600 B.C., depending on location.

The element iron was given its name by the Anglo-Saxons. It is a transition metal and today is mainly used to make steel. Alchemists used iron in many rituals, spells and amulets to stir and promote; energy, courage, strength, power, speed, determination, will-power, fertility-rites, assertiveness and aggression.

Coal (charcoal) was given a symbol often used for transformation, since during the process of burning matter it is transformed into warmth and ashes. Charcoal was often used by alchemists in purification processes and was said to represent personal vision, clarity and the release of emotional, physical and spiritual toxins. People have been making charcoal for thousands of years in China, West Asia, American, Africa and Europe.

It was highly prized by the ancient Egyptians as early as 1550 BC. The Japanese used bamboo and other charcoals for centuries in spiritual and physical healing, because it is able to absorb impurities and toxins. They believed that quartz crystal combined with charcoal produced a type of neutralising, positive ions in waveforms that would 'makes everyday life better'. It was considered an antidote for poisons in the environment and the body. Charcoal was used to help uplift and remove negative states of being and dissipate confusion to achieve balance. It was thought to bring about hope, spiritual psychic-abilities, true unconditional-love, tenderness, sensitivity, kindness and the appreciation of all things feminine in nature.

The preparation of charcoal for medical applications has been carried out for thousands of years. The ancient Egyptians used it for the adsorption of bad odour from decaying wounds. They also used charcoal in the preparation of papyri, which they used like paper as early as 1500 B.C. Hindu texts from 450 BC record the use of charcoal and sand filters for the purification of drinking water.

Charcoal residue consists of carbon and ash and can be made by removing water from burnt animals and vegetation, but it was usually produced by the slow burning of wood in the absence of oxygen. Generally it was made by piling wood up, covering it with dampened dirt and then the wood was set on fire. The process involved an irreversible altered state of chemical composition called Pyrolysis, a Greek-derived word from pyro meaning 'fire' and lysis which means 'separating'.

Although charcoal was used by many cultures as an art medium, it also played a major role in the development of mankind. First used as fuel for heat and later it enabled people to smelt and work metals. The production of charcoal and metallurgy exited side by side.

Potash, known today as Potassium carbonate is another substance used in alchemical processes. Most often it was symbolised by a plain rectangle standing on top of a cross. Alchemists retrieved this product from ashes produced by the burning of plants. The name refers to plant ashes, soaked in water in a pot. It produces a salt mixture that contains potassium in a water-soluble form.

Potash has been used since 500 A.D. in the process of making glass, the production of soap and for bleaching textiles. Nitrate of potassium used to be called saltpetre. Saltpetre was made from urine mixed with an organic matter such as straw. The ancient Chinese first used it to make fireworks and later European alchemist pyrotech's (from the Greek words 'pyro' meaning 'fire' and 'tekhnikos' meaning 'made by art') used it in the production of gunpowder.

8 **Bismuth** is another element that has been used in alchemy. There isn't a great deal of information about what sort of role it played in alchemical processes, but the element is known to have been used since the very early ages. Bismuth was first listed as an element by an unknown alchemist around 1400 AD. Later that century it was alloyed with lead to make cast type for printing and decorated caskets were being crafted out of this metal. It was represented by a symbol that looks like an eight with an opening at the top and as it has similar physical properties to lead and tin, it was often confused with these two elements until the 18th century. In chemical structure terms it is more akin to arsenic and antimony.

There are many more elements with allocated symbols used in alchemy and they were often laid out in a table format. A table that has been reshuffled, regrouped and added to over the last 300 years. It is now known as the Periodic Table. So Alchemy became Chemistry and Magic became Science.

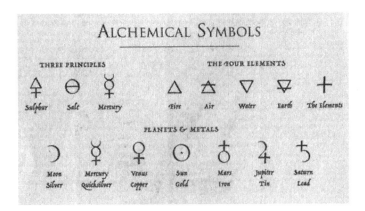

Basic Table of Alchemy Symbols

Alchemists like Newton would combine and mix different combinations of alchemy ingredients in the search for new and interesting reactions. They would then feverously report their findings to fellow alchemists, in order to gain credit and recognition. Each one secretly on a quest to make gold and ensure eternal life. These future Chemists would experiment and analyse the properties of each individual element, to discover their strengths and weaknesses. Listing and recording results in an orderly fashion but all behind closed doors, for fear of being accused of witchcraft.

For long periods of time Newton was obsessed with the

possibilities of alchemy and one of his servants recalled, "he very rarely went to bed until two or three of the clock, sometimes not till five or six, lying about four or five hours, especially at Springtime or Autumn, at which time he used to employ about six weeks in his laboratory, the fire scarce going out night or day. What his aim might be I was unable to penetrate into". We now know that all those long nights and focused days were taken up by Newton's experiments with alchemy, searching for the Philosophers Stone.

Newton's notebooks from 1662 exposes the development of Newton's intellectual interests. His practical projects at school and time spent with Mr Clark later helped him to formulate and carry out experiments unassisted and to build most of his scientific apparatus himself. The Pierpont Morgan notebook is filled with methods and ingredients for making paints and medicines, along with instructions for performing conjuring tricks. Later in the book there are lists of expenses that are filled with purchases of alchemical ingredients, as well as books and apparatus to equip the private laboratory he created in the grounds of Trinity College.

Alchemy as a subject was initially introduced to the world by Jabir ibn Hayyan (Geber), who was born in 721A.D. in Iran and died 813A.D. in Iraq, aged 94. He was the first person who considered the world around him from an alternative, alchemical perspective. Often regarded as the 'Father of Chemistry', he developed his own scientific and experimental approach to alchemy. By performing experiments and then recording the methods and results, Jabir ibn and other alchemists set down the foundations for modern chemistry.

The distinction between alchemy and chemistry started to emerge when a clear a line was drawn by Robert Boyle in his book, The Sceptical Chymist (Chemist) in 1661. Boyle's work questions the alchemic view of the four elements (earth, air, fire and water) and also the three primes (salt, sulphur and mercury). He had studied research carried out by Philippus Aureolus Theophrastus Bombastus von Hohenheim (what a splendid name), more commonly known as Paracelsus. He was born in 1439 and was the first person, in recent history, known for the common use of minerals and other elements in medicine, such as mercury, lead, arsenic and antimony. He was undoubtably responsible for many deaths as he was convinced that what we now regard as poisons, were cures for many ails. It is said that he even attended Louis XIV as his senior medical advisor. Boyle keenly studied Paracelsus's doctrines about elements and principles, some of which were actually good and he laid down his own new, vastly improved set of definitions.

Robert Boyle was the first person for centuries to conclude that matter consists of atoms. He also later realised that chemical elements came in families and could be organised into groups. Chemists would eventually understand that this pattern is to do with the way electrons organise themselves in orbits around atomic nuclei.

Newton didn't publish any of the vast amount of work he carried out during his alchemy pursuits. Partly because he was rather secretive about much of his work but also because he thought that he was making progress towards alchemy's greatest challenge, the Philosophers Stone. So he endeavored to keep his potential gold making method to himself. Further fear came from the fact that alchemy had been made illegal prior to Newton's birth and there still remained a damning stigma against its practice. In addition, the actual practice of trying to make gold or silver was a criminal offence and had been since a law had been passed by Henry IV in 1404. In spite of this, throughout his life he and other scientists of the time, many of whom were fellows of the Royal Society, carried out extensive research into alchemy.

There are some documents which emerged long after Newton's death full of alchemist research. They are generally written in English, but it is difficult to read and understand what Newton was actually saying in the few documents found. Alchemists were notorious for recording their methods and theories using their symbolic language or code, so others could not understand their work. They did this in order to protect themselves from possible repercussions. Extra secrecy was required by Newton as his reputation grew and awards and rewards became aplenty, as he had more and more to lose by exposing his secret beliefs.

In the days before the venomous disagreements between

Newton and Boyle, the two men were amiable colleagues. On one occasion Newton wrote to his fellow alchemist Robert Boyle requesting him to maintain "high silence" in openly divulging the principles of alchemy. Newton continued "because the way by the Mercurial principle may be impregnated has been thought fit to be concealed by others that have known it and therefore may possibly be an inlet to something more noble that is not to be communicated without immense damage to the world if there be any verity in".

This concern for the world in relation to alchemy was mirrored by others and even though it is thought he wrote over a million words relating to alchemy, following his death the Royal Society declared his works in this field as "not fit to be printed". Newton's papers on the subject vanished from view for a long time due to a concern that his interest in alchemy might, in some way, devalue or discredit his other work that the Royal Society had applauded.

It is hardly surprising that the Royal Society took this stance as although during his alchemical pursuits Newton applied his usual rigorous scientific methods, by carrying out many meticulous experiments and recording lengthy notations of his findings. It is hard to believe that the author of the Principia also spent so much time writing strange and decidedly unscientific methods such as –

"Dissolve volatile green lion in the central salt of Venus and distil. This spirit is the green lion the blood of the green lion Venus, the Babylonian Dragon that kills everything with its poison, but conquered by being assuaged by the Doves of Diana, it is the Bond of Mercury".

The papers resurfaced in the middle of the twentieth century and the general opinion of the majority of academics is that Newton was first and foremost an alchemist. Conceding that the inspiration for Newton's theory of gravity and laws of light emerged from his alchemical work. He did not publish any of his work on alchemy. Some say that this should not infer that he considered that he had failed in his main pursuit of the Philosophers Stone. They have concluded that it shows his success was great enough to convince him that he was right and his quest was possible, giving him good cause to maintain his "high silence". When considering his position as Master of the Royal Mint and his attitude to the forgery of gold coins, his dilemma is apparent and provides an excellent standalone reason for keeping quiet about his endeavours.

Others that are certain that much of Newtons research and belief in alchemy didn't really amount to much, point out that he never did manage to actually turn lead into gold. But he was right to believe that it was possible. As it transpired three centuries later, using nuclear transmutation it has been found possible and easier to turn gold into lead than the reverse reaction. However, in 1932 nuclear experiments successfully transmuted lead into gold, using a particle accelerator, or atom smasher. By way of the reaction of fissionable isotopes in a nuclear reactor and by the bombardment of nuclei with the accelerator, using high speed particles. This caused a change in the nucleus, or core of the atom and was the very first true transmutation. So far the expense of the procedure far exceeds any gain. But Newton was correct in his hypothesis, even if his method was missing a nuclear reaction. He knew it was possible and spent a massive portion of his life trying to prove it.

Alchemical teachings and research encouraged Newton in many directions and may have inspired his discovery that white light is a mixture of various colours. Alchemists were the first to work out that compounds can be broken down into their constituent parts and then recombined. Newton adapted and applied that science to white light, which he deconstructed into each separate constituent colour and then recombined. Alchemy had given Newton an opening in his mind that allowed him to see that what was thought of as impossible, might actually be possible. It allowed him to see beyond logic. Therefore, it could be said that if it hadn't been for Newton the alchemist, we might not have learned some of the most famous discoveries from Newton the scientist.

Many of Newton's texts relating to alchemy may have been destroyed. Others are so full of strange unknown symbols, ambiguous code and inexplicable details that it has been impossible to decipher. But some of his surviving texts are now available for study and give an interesting insight into the unique mind of a complicated, mystically inclined, highly ambitious genius, who was determined to understand the Universe in any and every possible way.

"Errors are not in the art but in the artificers".

Isaac Newton

Artificer definition - a person who is skilful or clever in devising ways of making things; inventor.

Chapter 7

What Newton Got Wrong

Because Newtons mind was so open to impossibilities, he sometimes roamed off the tracks of reality. The contradiction in his nature can be seen by his leaps to unbelievable conclusions and yet he was such an exacting man. A man who made very few mistakes. Even when he did make mistakes, it was usually due to the technology constraints of the day. Just think what he could have done if electricity had been available to him. Most often he worked things out eventually, but not always. The thing to remember is that nobody gets it right all the time and the main mission of the scientist is to keep researching, recording and experimenting, until they find the solution or the truth that they seek.

Aspects of Gravity

At one time Newton was convinced that the term gravity indicated some built in property produced by the material an object was made of, instead of an external force exerted by the Earth. Which he later came to realise was the truth. Once he had a better understanding of gravity, he correctly explained its existence and the effect it has on our lives, but he was wrong about what created gravity. He had concluded that gravity was some sort of universal pulling force that held everything together and that it was just part of the way things worked on Earth and in the Universe. More recently Albert Einstein explained that gravity is more than just a force. It is a curvature in the spacetime continuum. The mass of an object causes the space around the object to bend and curve. According to Albert Einstein's theory, the reason mass is proportional to gravity is because all things with mass emit minute particles called gravitons, that are responsible for gravitational attraction. Therefore the more mass an object has, the more gravitons there will be, which exist like an invisible cloud around an object.

Specifics of Light

Newton's work and observations led him to a theory of colour and light, which he had worked on between 1666 and 1670. His final conclusions showed that he believed that light was made up of rays of minute 'corpuscles', the size of atoms. It is now known that the details of this particular theory are incorrect. The notion of light corpuscles would be investigated a few hundred years later by Einstein. These days we call corpuscles 'photons', however Einstein's light corpuscles are quantum particles, which do not obey Newton's laws.

The Cat Flap

Not everyone is convinced that this is true, however the story goes that Newton found himself being constantly interrupted during his intense periods of study by his cats pestering to be let out. He also found the shadows of the cats interfered with the light coming from the oil lamps when he was working deep into the dark nights, at his Cambridge laboratory. In utter frustration at the constant scratching and meowing he endeavoured to do something to prevent it. According to Wikipedia, the invention of the household pet door was attributed to Isaac Newton in a story authored anonymously and published in a column of anecdotes in 1893. To the effect that Newton foolishly made a large hole for his adult cat and a small one for her kittens, not realising that the kittens would follow the mother through the large one. Working with animals is always a tough one.

Atoms

Newton wrote in an unpublished paper that Empedocles, an ancient, pre-Socratic Greek philosopher who lived 490 BC - 434 BC, considered that all matter must consists of atoms. This Newton dismissed as "a very ancient opinion".

The Speed of Sound in Air

After performing the first analytical determination of the speed of sound in air, Newton published his findings in Book II of Principia Mathematica, as proposition 49. He believed that he could correctly predict the speed of sound though the different mediums of solids, liquids and gases, as long as he knew the density of the medium and the pressure acting on the sound wave. He decided that he could work out the speed of sound by calculating the square root of the pressure, divided by the medium's density.

He wasn't too far wrong in his theory. For sea level air at a typical ambient temperature he calculated a value of 979 ft per second, which is too low by about 15%. The true value being about 1116 ft per second.

Fuel for the Sun

In conversations that Newton had with his Nieces husband John Conduitt towards the end of his life, Newton explained his beliefs about how the sun is fuelled. He said –

"Comets, in looping past the Sun, slowly become cooked. The Sun would go out due to its constant conflagration, but fortunately, it is constantly renewed and powered by the fresh fuel provided it by comets passing nearby. Intelligent beings (angels or aliens) oversee this whole process of regular fuelling of the Sun, but unfortunately the Great Comet of 1680 passed so close to the Sun that it seems likely that it will soon fall directly into the sun, rather than feeding it a small measure of fuel. With an enormous quantity of fuel abruptly dumped into the Sun instead of dispensed over eons, it will flare up like a bonfire and quite likely roast the Earth (possibly as a Red Supergiant might), killing everything on it. This possibly happens regularly and after this, possibly God would renew creation by repopulating instead the moons of Saturn or Jupiter which would have escaped the inferno relatively intact due to their distance".

We now know that the vast majority of that particular theory is completely untrue. Yet at the time, in the early 1700's, most of the great physicists, mathematicians, astronomers, natural philosophers, alchemists and theologians hung on to Newtons every word. However, regardless of the fact the Principia is one of the most influential books in the history of science, which dominated the scientific view of the physical Universe for the following three centuries. In spite of the fact that Newton remains a giant in our minds on a par with geniuses such as Archimedes and Gauss, sometimes his thoughts were utterly bonkers.

Human Existence
Being such a reverent religious man, Newton studied ancient theological works most passionately. As a result of his delving and curious disposition, his main focus and motivation was that he wanted to discover a single system of the world that would explain everything. One result of his study was that Newton seemed to have been convinced that the human race had been on Earth for a much shorter time than is currently believed. His writings imply, quite bizarrely that humanity has existed for a long time but in perpetual rotations of existence.

Followed by long periods of virtual extinction. Bringing the possibility of unlimited rotations of knowledge, intended for, or likely to be understood by only a small number of people. Knowledge that must only be given to those with a specialised understanding or interest, who would see the secret messages that had been ciphered into ancient and religious writings. He believed he was following up on information left by ancient civilisations such as the Mayans and Mesoamericans and also clues hidden in ancient texts such as the Hebrew Bible and the old Testament.

There are gaps in the history of the World and after Newton pondered these gaps he concluded, "after billions of years, humans evolved, but then nothing of any importance happened for hundreds of thousands or millions of years. Until then everything started happening in the past 10,000 years, at an ever-accelerating pace".

This idea and line of thought can be related to alchemy and his biblical studies, but Newton did not come up with the theory of an edited human history. It was first raised by Lucretius's, who wrote 'On the Nature of Things' 50 BC.

Titus Lucretius was a Roman philosopher and poet and he wrote -

"Moreover, if heaven and earth never had a beginning or birth, but have existed from everlasting, why have there not been other poets to sing of other events prior to the Theban war and the tragedy of Troy? Why have so many heroic deeds so often been buried in oblivion, instead of flowering somewhere, implanted in eternal memorials of fame? The true explanation, in my judgment, is that our world is in its youth, it was not created long ago, but is of comparatively recent origin. That is why at the present time some arts are still being refined, still being developed. This age has seen many improvements in shipbuilding; it is not long since musicians first moulded melodious tunes; our system of philosophy too is a recent invention and I myself am found to be the very first with the ability to expound it in the language of my country [Latin].

If by chance you believe that all these same things happened before, but that the races of human beings perished in a great conflagration, or that their cities were razed by a mighty convulsion of the world, or that rivers, rapacious after unremitting rains, inundated the earth and submerged towns, there is all the more necessity for you to admit defeat and acknowledge that heaven and Earth are destined to be destroyed".   Titus Lucretius

With current weather conditions, global warming, flooding, sea levels rising as never before, just in the last century. Bush fires often consuming vast areas of the world's forests and woodlands, earthquakes and other violent movements of the Earth's crust, such as recent tsunami's. That final paragraph by Lucretius's is a fair description of what is happening to the world today. Are we now destined for another phase of virtual extinction followed by the rebirth of our race. Was Newton wrong to believe this theory on life or does it explain why the Mayans with their vast nation simply vanished from history. Give true reason for the end of the Pharos dynasties, which inexplicably ceased to exist without a single hieroglyph explaining the cause.

Either way, as Newton's prediction that the end of the world would be no sooner than the year 2060, for now, life goes on. This was a Bible-based prediction for the end of the world founded on Newton's mathematical analysis of the Bible. This wasn't the only calculated event Newton claimed he extracted from the Bible, he also insisted that his work led him to believe that he knew the precise dates for the creation of Noah's ark and other random biblical events.

Isaac Newton was
very creative and
loved to draw and
paint. It was said that
he covered the walls
and ceiling of his
bedroom with different
colours.

"What goes up,

must come down".

Isaac Newton

Chapter 8

Newton Inspired Future Scientists & Innovation

Newton once said –

"If I have seen further, it is by standing on the shoulders of Giants".

Since his death he has become one of those giants and many scientists have seen further than Newton ever did, by standing on his shoulders and taking one, or many steps further.

Sometime after Newtons death, the 1730s and 40s observed a vast overhaul of the claims and stability of the science in the Principia, which presented a number of unanswered loose-ends. Questions that only other scientists would notice had been left without conclusions. Some even dared to question just how true Newton's theory of gravity actually was. The Royal Society had been pushed to conduct expeditions to Peru and Lapland, to prove once and for all whether Newton's claims about the non-spherical shape of the Earth and also the variation of the surface gravity with latitude, were actually correct. Some of Newton's claims were successfully confirmed as true in the 1740's, following the return of the expedition from Peru. The French physicist and mathematician Jean Le Rond d'Alembert's 1749 rigid-body solution for the wobble of the Earth, agreed that it is the non-spherical shape of the Earth which produces the annual seasons as the precession of the equinoxes.

Nikola Tesla said he was greatly inspired by one of Newton's quotes. When Newton was asked how he came up with solutions to such complex questions, he said -

"I simply hold the thought steadily in my mind's eye until a clear light dawns upon me".

Tesla practiced this method when he came to a halt in his thinking and produced some of the most brilliant ideas and inventions known to human kind. What a thought, a genius, inspiring a genius. Without such inspiration, we might not be enjoying all that Tesla brought to the world, including AC electric supplies that power all our gadgets today.

Albert Einstein was greatly influenced by Isaac Newton. He considered him to be 'the most-genius physicist' and said Newton inspired him to look further. Einstein realised that Newtons knowledge of gravity was not fully explored. A few hundred years had passed since Newton had first realised the presence of gravity and its effects on people, the planets and everything else in the Universe. Einstein went on to develop his General Theory Of Relativity to finally explain the vital missing aspects of Newton's theory of gravity.

Industrial Revolution

Newton's book the Principia and his Three Laws of Motion ultimately induced the Industrial Revolution in England, that occurred during a period from 1760 to 1840. Newton's discoveries were instrumental in enabling the transition to new, improved manufacturing processes in Europe. The transformation included a change from hand production methods to machine made products. Machines that operated according to Newton's brilliantly defined laws. New factories and locomotives were built according to the great man's science creating an increased use of steam and water power plus the development of machine tools.

"Truth is ever to be found
in simplicity and not in
the multiplicity and
confusion of things".

Isaac Newton

Chapter 9

Newton's Timeline

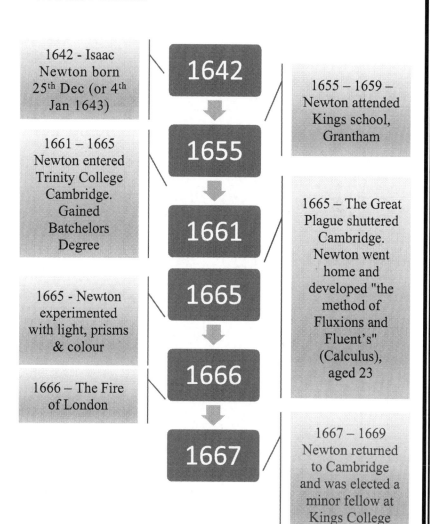

1642 - Isaac Newton born 25th Dec (or 4th Jan 1643)

1661 – 1665 Newton entered Trinity College Cambridge. Gained Batchelors Degree

1665 - Newton experimented with light, prisms & colour

1666 – The Fire of London

1642

1655

1661

1665

1666

1667

1655 – 1659 – Newton attended Kings school, Grantham

1665 – The Great Plague shuttered Cambridge. Newton went home and developed "the method of Fluxions and Fluent's" (Calculus), aged 23

1667 – 1669 Newton returned to Cambridge and was elected a minor fellow at Kings College

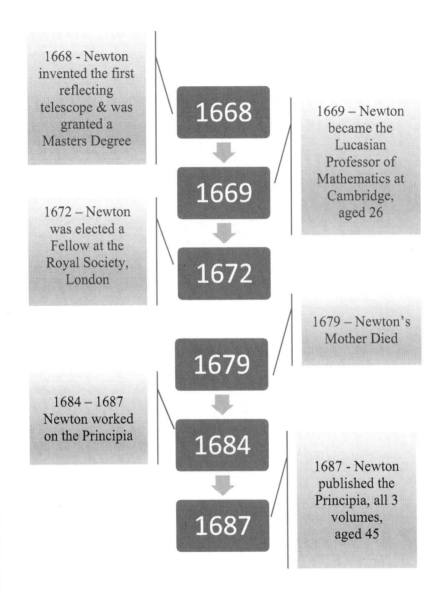

1668 - Newton invented the first reflecting telescope & was granted a Masters Degree

1668

1669 – Newton became the Lucasian Professor of Mathematics at Cambridge, aged 26

1669

1672 – Newton was elected a Fellow at the Royal Society, London

1672

1679 – Newton's Mother Died

1679

1684 – 1687 Newton worked on the Principia

1684

1687 - Newton published the Principia, all 3 volumes, aged 45

1687

1689 – 1690
Newton served as a member of Parliament in England

1689

1693 – Newton suffered an emotional, nervous breakdown, aged 51

1693

1696 – Newton became Warden of the Royal Mint, moved to London & almost completely stopped lecturing at Cambridge

1696

1699 – Newton was elected to the Council of the Royal Society

1699

1700 until his death - Newton was appointed as Master of the Royal Mint.

1700

1701 - 1702 Newton served again as a Member of Parliament in England. He also resigned the post of Lucasian Professor of Mathematics at Cambridge

1701

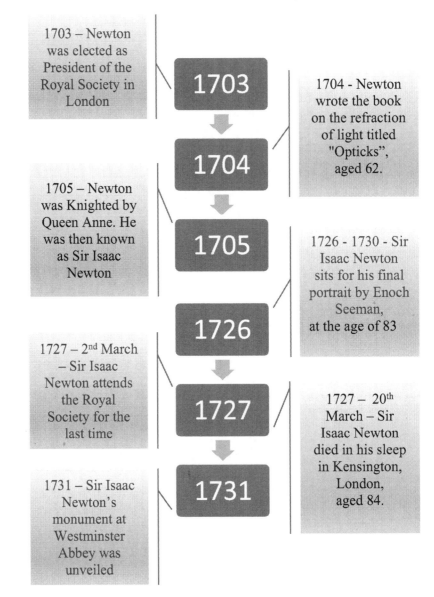

1703 – Newton was elected as President of the Royal Society in London

1703

1704 - Newton wrote the book on the refraction of light titled "Opticks", aged 62.

1704

1705 – Newton was Knighted by Queen Anne. He was then known as Sir Isaac Newton

1705

1726 - 1730 - Sir Isaac Newton sits for his final portrait by Enoch Seeman, at the age of 83

1726

1727 – 2nd March – Sir Isaac Newton attends the Royal Society for the last time

1727

1727 – 20th March – Sir Isaac Newton died in his sleep in Kensington, London, aged 84.

1731 – Sir Isaac Newton's monument at Westminster Abbey was unveiled

1731

"If I have ever made any valuable discoveries, it has been owing more to patient attention, than to any other talent".

Isaac Newton

Chapter 10

Activities, Tricks & Things to Do That

Demonstrate Newtons Laws, Forces & Discoveries

Reminder

Sir Isaac Newton developed Three Laws of Motion. His laws of motion are three physical laws that together laid down the foundations for classical mechanics. They neatly describe the relationship between a body and the forces acting upon it and its motion in response to those forces. They are –

1ˢᵗ Law = Inertia, which says that an object's speed will not change unless something makes it change.

2ⁿᵈ Law = Action-reaction, which states that the strength of the force on an object is equal to the mass of the object, times the resulting acceleration.

3ʳᵈ Law = Force, The third law says that when two objects interact, they apply forces upon each other of equal strength but in opposite directions.

To understand and take in these somewhat complex laws by just memorising the words does not really bring to mind how they actually work in reality. The next few activities aim to demonstrate these laws to provide a better understanding of them.

Activity 1 – Demonstrates Newton's 1st Law - Inertia

The Law of Inertia - A body at rest remains at rest and a body in motion remains in motion with a constant speed and in a straight line unless acted upon by an outside force.

For this trick you need –

- a bottle, with the lid removed
- a new ten pound note
- some coins (3 pound coins will do)

Place the note on the mouth of the bottle (try using a plastic bottle at first, but a glass bottle is better when your well-practiced). Offset the note slightly so one side is a bit longer for easy practice.

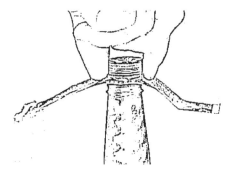

Place the coins on top of the note over the mouth of the bottle.

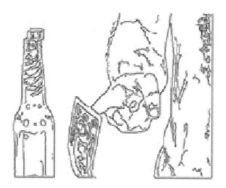

Point your index finger at the overhanging note and use a quick Karate chop on the note, without touching the bottle or the coins.

The note comes out and the coins remain balanced on the top of the bottle.

Science Behind the Trick

This trick works because of inertia. Newtons first law of motion states that inertia is the tendency of an object at rest to remain at rest, unless an outside force acts upon it and is strong enough to overcome the objects inertia, then the object will move.

Inertia is visible in the bottle trick as the bottle and coins do not move and won't move unless an outside force, moves them.

Friction, or the lack of it also plays an important role in this trick. Friction is the tendency of one surface to grab or cling to another surface moving across it. The surface of the new note is clean and smooth. It doesn't create a lot of friction against the stack of coins above it or the bottle below it. Without friction, the note doesn't pull the coins or move the bottle and they remain balanced on the top of the bottle mouth. This trick wouldn't, for example, work with sandpaper which produces a great deal of friction.

Expand the trick -

- Test different types and thicknesses of plastic, paper or card to identify the best possible material for the trick.
- Establish ways to change the friction between the note and the objects that sandwich it. Try changing the bottle to plastic or tin and the coins to 2p's or 50p's. Which works best?

- Look at how different masses (weights) of the stacked objects change the outcome by using more and more coins. What is the maximum number you can successfully do the trick with?
- Will an empty bottle give a different outcome to a full bottle?
- Lubricant can be used to reduce friction between two surfaces moving across each other. Lubricants are not always liquid. What materials can you find that can act as a lubricant? For example, sand can give movement.

These are just a few ideas. You can come up with your own variables to test. You can change just one variable for each test while making sure that all the other factors in your test remain the same. Then you can scientifically compare and analyse your results.

Be careful ! - It should be obvious that practicing with a plastic bottle is a lot safer than with glass. Also keep in mind that objects with more weight will hurt your toes more, should they fall, than those with less weight.

Activity 2 – Demonstrates Newton's Law of Inertia & Gravity

So remember, a body at rest remains at rest and a body in motion remains in motion with a constant speed and in a straight line unless acted upon by an outside force.

Nut in a Bottle -

For this trick you need – a bottle

- a nut from a bolt

- a small plastic hoop

Place the bottle on a firm flat surface. Then balance the hoop on top of the bottle and balance the nut on top of the hoop. Be sure to check that the nut is balanced directly over the bottle opening.

Then quickly snatch the ring away from the bottle and the nut will fall into the bottle.

Science behind the trick

The first of Newtons three laws of motion states that an object at rest will remain at rest unless something makes it move. By hitting the ring on its outside, you overcome its stationary inertia. The plastic flexes upward and that upward flex overcomes the nut's stationary inertia. This pushes the nut up and away from the bottle until gravity curves it back down towards the table and into the bottle.

However, when you hit the ring on its inside, the plastic flexes downward. The plastic actually drops out from under the nut and is pulled away sideways. The nut is no longer supported and because it's not moving due its stationary inertia, it doesn't do anything for a fraction of a second, until gravity overcomes the nut's stationary inertia and pulls the nut straight down into the bottle. Practice makes perfect with this trick but it's very satisfying when you've mastered it.

Note that, once the nut is moving it won't stop moving, speed up, slow down, or change directions, due to its moving inertia. That is unless something like your hand, the bottle or the floor gets in the way and causes the nut to stop or change direction.

Analyse the variables -
Knowing what you now know about inertia and gravity, test other variables.

- You can start by increasing the number of nuts to see if that changes the outcome. Make sure the nuts are piled neatly on top of each other horizontally as they

- all need to fall in the bottle as a stack, so try and keep them lined up.

- Try stacking the nuts vertically instead of horizontally. This gets more difficult the more nuts you use. You'll need a steady hand.

- Try different objects like a pile of 5p's, a new unsharpened pencil, or a marker with a flat end on it. What else can you think of to show inertia using this gravity drop method?

- Using your iPhone to take a slow-motion video of your experiments. Capturing your tests in slow motion will reveal some interesting details that you can't detect with your eyes.

Activity 3 – Demonstrates Newton's 3ʳᵈ Law of Motion

Remember - For every action there is an equal and opposite reaction.

This simple science experiment is such an easy and fascinating way to demonstrate how energy is transferred from one object to another and how an equal and opposite reaction occurs.

For this experiment you will need –

- Marbles - 5 of the same size.

-A long ruler with a indent/channel down the middle or any straight indent where the marbles can run smoothly.

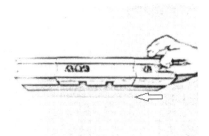

- Set the marbles on the ruler. One on its own at one end and three more half way down the channel.
- Flick the first marble quite quickly and firmly.

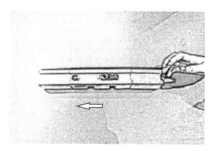

Interestingly, only one of the marbles rolls away.

- Repeat the experiment with different configurations of marbles such as shooting 2 marbles at 3.

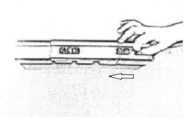

Soon a pattern emerges, the number of marbles that you flick equals the number of marbles that roll away.

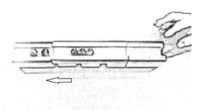

In other words, a rolling marble can only transfers its energy to the marble it collides with, which then transfers the same energy to the next marble and so on. Creating a series of equal and opposite reactions.

Note

In order to get the best results with this experiment, you will need to make sure that the surface you are working on is level. Gravity will cause the marbles to roll away if there is any slant to your table.

Also, you will need to give the marble a good, solid flick. Not too wild – you don't want marbles all over the place. But if your flick is too weak you will not achieve the desired results.

Activity 4 – Demonstrates Gravity, as defined by Newton

The Apple Drop -

This experiment helps give a basic understanding of how gravity operates.

For this experiment you will need –

- 1 x Whole apple

- 1x Apple cut in half

- 1 Chair

- iPhone with Slow-Motion Camera

Consider what you already think and believe about gravity. Look at the whole and the half apple, do you think the whole apple or the half apple will hit the ground first, if you drop them from the same height at the same time?

- Stand on a chair and drop the whole apple and the half apple at exactly the same time.
- Ask your friend to film the apples whilst they fall.
- Watch the film back in slow motion.

Both the whole and the half apple will land at the same time, because all objects descend at the same speed. It's difficult to believe as it seems logical that the heavier whole apple should reach the floor first. But if seeing isn't believing, try listening.

- Try dropping the apples again but this time close your eyes and listen to the thumps as the apples land at exactly the same time on the floor.
- Try the same experiment with 2 different objects and from different heights.

It seems illogical and yet it is the truth about gravity and Newton was the first to recognise this natural and yet strange force, that we all live under.

Activity 5 – Demonstrates the Orbits of the planets

And Other Stuff

For this activity you need – An iPhone

The iPhone 'Planets' app. Just download it from the App Store for free.

Also available for Android at : http://itunes.com/app/planets android.

Exercise 1

To give you a sense of your own planet and where exactly you are on it –

- On your iPhone download the free app called 'planets'. It offers a view of the universe in real time from exactly where you are right now.

- Tap on the 'Globe' tab. Identify which countries are enjoying daylight at the same time as you. The globe can be stopped from turning by tapping on the screen.

- Using the 'Options' tab make a note of your own particular longitude and latitude by turning on 'Automatic Location'.

- Then turn off 'Automatic Location', tap the 'Location' option and enter another city e.g. Sydney. You can find the longitude and latitude for any place on the planet.

By cross referencing the longitude and latitude on a map of the globe, you can identify your location anywhere on the planet. Very useful if your lost at sea.

Exercise 2

See how planet Earth compares with the other planets.

- Use the 'Visibility' tab to see when each planet is actually in the sky above you by looking at the times shown above each planets bar. Which planet is visible longest?

We already know it takes one year for the Earth to make a complete orbit of the Sun. We also know that the Moon travels once around the Earth every 24 hours. But what about the other planets in our Solar System.

- Tap on the arrow next to each planet to see more information. You can find out how long it takes the other planets to complete an orbit of the sun by tapping on each planet in turn on the 'Visibility' tab and looking down the list for the 'Orbital Period' readings.

- Find and write down the 'Orbital Period' for each planet. Which planet takes 164.8 years to orbit the Sun?

- You can also see on each list which of the planets are classified as Terrestrial, Gas Giants or Ice Giants.

- Check out how many moons each planet has? Some planets have no moons. But which one, astonishingly has 67 moons?

- You can go further and see what the atmosphere of each planet is made up from in chemical terms.

- Look at the 'Mass' and 'Radius' of each planet. This will give you a sense their size in comparison with Earth.

- The 'Sky 3D' tab lets you see what the Sun, the Moon and planets are doing beyond the clouds, even in daylight. It shows which planets are rising in real time as well as all the constellations and their names.

- By tapping the 'Visible' option at the top of the screen on the 'Sky 3D' tab, you can view gamma rays, x-rays and other occurrences in the solar system that we can't see with our eyes. Along with many other exciting options.

Newtons laws made sense of our Solar System, explained the Seasons and even found the planet Neptune.

Activity 6 – Demonstrates how white light can be separated into the colours of the rainbow.

Make a rainbow and see how it works

For this activity you will need –

- A glass of water ¾ full

- White paper

- A sunny day

Method –

1) Take the glass of water and white paper to an area with sunlight, near a window.

2) Hold the glass of water, without spilling it, above the paper and watch as the sunlight passes through the glass of water, refracts (bends) and forms a rainbow of colours on your white paper.

3) Try holding the glass of water at different heights and angles to see if it has different effects.

Science behind the Rainbow

Whilst normally a rainbow can be seen as an arc of colour in the sky, they can also be formed in other situations. You may have seen a rainbow in a water fountain or in the mist of a waterfall and you can make your own as you did in this activity. Rainbows form in the sky when sunlight refracts (bends) as it passes through raindrops, it acts in the same way when it passes through the glass of water. The sunlight refracts, separating the white light into the colours red, orange, yellow, green, blue, indigo and Violet.

If you have access to two prisms, you could actually recreate Newtons experiment where he separated light into individual colours with one prism and then converted them back to white light with the second on.

Activity 7 – Explains Newton's Law of Cooling

For those of you who enjoy a mathematical challenge the following activity explains in detail Newton's Law of Cooling and explains his formula. This activity is not to test you, but more to demonstrate what a brilliant mathematician Newton was. But if you want to understand it, you need to do it.

For this activity you will need –

- A pen

- Your brain

- Probably a calculator

- Logarithms

1) A pan of soup starts at a temperature of 373.0 K.
2) The surrounding temperature is 293.0 K
3) The cooling constant is $k = 0.00150$ $1/s$

Question 1 - what will the temperature of the pan of soup be after 20.0 minutes?

Method - The soup cools for 20.0 minutes, which is:

$$t = 20.0 \; min \times \frac{60 \; s}{1 \; min}$$

$t = 1200 \; s$

Newton's formula says –

The temperature of the soup after the given time can be found using the formula:

$$T(t) = T_s + (T_0 - T_s) \; e^{(-kt)}$$

$T(1200 \; s) = 293.0 \; K + (373.0 \; K - 293.0 \; K) \; e^{(-(0.001500 \; 1/s)(1200 \; s))}$

$T(1200 \; s) = 293.0 \; K + (373.0 \; K - 293.0 \; K) \; e^{(-1.800)}$

$T(1200 \; s) = 293.0 \; K + (373.0 \; K - 293.0 \; K)(0.1653)$

$T(1200 \; s) = 293.0 \; K + (80.0 \; K)(0.1653)$

The final steps and the answer can be found on the back page of this book. Now try the next one -

1) A rod of iron is heated in a forge to a temperature of 1280.0K.

2) The rod is then plunged in to a bucket of chilled water with a temperature of 280.0 K.

3) After 10.0 s, the temperature of the iron rod drops to 329.7K.

Question 2 - What is the cooling constant for this iron rod in water?

Method - The cooling constant can be found by rearranging the formula:

$T(t) = T_s + (T_0 - T_s) e^{(-kt)}$

$\therefore T(t) - T_s = (T_0 - T_s) e^{(-kt)}$

$$\therefore \frac{T(t) - T_s}{T_0 - T_s} = e^{-kt}$$

The next step uses the properties of logarithms. The natural logarithm of a value is related to the exponential function (e^x) in the following way: if $y = e^x$, then $\ln y = x$. To continue rearranging the formula, the natural logarithm is applied to both sides.

$$\frac{T(t) - T_s}{T_0 - T_s} = e^{-kt}$$

$$\therefore \ln\left(\frac{T(t) - T_s}{T_0 - T_s}\right) = \ln(e^{-kt})$$

$$\therefore \ln\left(\frac{T(t) - T_s}{T_0 - T_s}\right) = -kt$$

$$\therefore -kt = \ln\left(\frac{T(t) - T_s}{T_0 - T_s}\right)$$

$$\therefore k = -\frac{1}{t}\ln\left(\frac{T(t) - T_s}{T_0 - T_s}\right)$$

Using this formula for the cooling constant k, the solution can be found:

$$k = -\frac{1}{t}\ln\left(\frac{T(t) - T_s}{T_0 - T_s}\right)$$

$$k = -\frac{1}{10.0\ s}\ln\left(\frac{(329.7K) - (280.0K)}{(1280.0K) - (280.0K)}\right)$$

$$k = -\frac{1}{10.0\ s}\ln\left(\frac{(49.7\ K)}{(1000.0\ K)}\right)$$

$$k = -\frac{1}{10.0\ s}\ln(0.0497)$$

The final steps and the answer can be found on the back page of this book.

Activity 8 – Alchemy Spells

For this activity you will need –

- This book

- Google

- No cauldron required

Look up the symbols below, in the 'Newton & Alchemy' chapter in this book to find which elements they represent. Then use Google to search for at least two ancient or modern applications for their use.

Spell 1

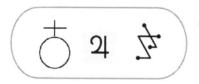

Spell 2

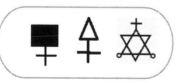

- Try making your own alchemy spells from the elements available in Newton's day.

- Then see what spells you can create with all the elements that have been identified to date, shown on the periodic table, which you can look up on Google.

Books by Isaac Newton - Published during his lifetime

De analysi per aequationes numero terminorum infinitas (1669, published 1711)

Method of Fluxions (1671)

Of Natures Obvious Laws & Processes in Vegetation (unpublished, c. 1671–75) [148]

De motu corporum in gyrum (1684)

Philosophiae Naturalis Principia Mathematica (1687)

Opticks (1704)

Reports as Master of the Mint (1701–25)

Arithmetica Universalis (1707)

Newton Books published posthumously

The System of the World (1728)

Optical Lectures (1728)

The Chronology of Ancient Kingdoms Amended (1728)

De mundi systemate (1728)

Observations on Daniel and The Apocalypse of St. John (1733)

Newton, Isaac (1991). Robinson, Arthur B., ed. Observations upon the Prophecies of Daniel, and the Apocalypse of St John. Cave Junction, Oregon: Oregon Institute of Science & Medicine. ISBN0-942487-02-8. (A facsimile edition of the 1733 work.)

A Historical Account of Two Notable Corruptions of Scripture (1754)

References

Annan, Noel, The Dons: Meteors, Eccentrics, and Geniuses. London: Harper Collins, 2000.

Asimov, Isaac, The History of Physics. New York: Walker & Co., 1966.

Brock, William H., The Norton History of Chemistry. London: W.W. Norton, 1993.

Carlson, Bernard W., Tesla: Inventor of the Electrical Age. Princeton: Princeton University Press, 2013.

Christianson, Gale E., In the Presence of the Creator: Isaac Newton and his Times. New York: Free Press/ Macmillan, 1984.

Cockren, Archibald, Alchemy Rediscovered and Restored. Philadelphia: David McKay, 1941.

Cohen, I. B. and Smith, G. E., The Cambridge Companion to Newton, Cambridge: Cambridge University Press,2002.

Cole, K. C., First You Build a Cloud: And Other Reflections on Physics as a Way of Life. San Diego: Harvest/Hardcourt Brace, 1999.

Cox, Brian and Jeff Forshaw, Universal: A Guide to the Cosmos. London: Penguin Books, 2016

Cropper, William H., Great Physicists: The Life and Times of Leading Physicists from Galileo to Hawking. Oxford: Oxford University Press, 2002.

Cohen, I. B. and Westfall, R. S., Newton: Texts, Backgrounds, and Commentaries, A Norton Critical Edition, New York: Norton. 1995

Dobbs, B.J.T., The Foundations of Newton's Alchemy. Cambridge University Press, 1984.

Ebbing, Darrell D., General Chemistry. Boston: Houghton Mifflin, 1996.

Feingold, Mordechai, 2004, The Newtonian Moment: Isaac Newton and the Making of Modern Culture, Oxford: Oxford University Press, 2004.

Ferguson, Kitty, Measuring the Universe: The Historical Quest to Quantify Space. London: Headline, 1999.

Ferris, Timothy, The Mind's Sky: Human Intelligence in a Cosmic Context. New York: Bantam Books, 1992.

Ferris, Timothy, The Whole Shebang: A State of the Universe(s) Report. London: Phoenix, 1998.

Giancola, Douglas C., Physics: Principles with Applications. London: Prentice-Hall, 1997.

Gribbin, John, Almost Everyone's Guide to Science: The Universe, Life and Everything. London: Phoenix, 1998.

Hall, A. Rupert, Isaac Newton: Adventurer in Thought, Oxford: Blackwell. 1992.

Harrison, Edward, Darkness at Night: A Riddle of the Universe. Cambridge, mass.: Harvard University Press, 1987.

Hartmann, William K., The History of Earth: An Illustrated Chronicle of an Evolving Planet. London: Workman Publishing, 1991.

Hawking, Stephen, A Brief History of Time: From the Big Bang to Black Holes. London: Bantam Books, 1988.

Hawking, Stephen, The Universe in a Nutshell. London: Bantam Press, 2001.

Heiserman, David L., Exploring Chemical Elements and their Compounds. Blue Ridge Summit, Pa.: TAB Books/ McGraw Hill, 1992.

Holmyard, E. J., Makers of Chemistry. Oxford: Clarendon Press, 1931.

Krebs, Robert E., The History and Use of our Earth's Chemical Elements. Westport, Conn.: Greenwood, 1998.

Leicester, Henry M., The Historical Background of Chemistry. New York: Dover, 1971.

Lewis, John s., Rain of Iron and Ice: The Very Real Threat of Comet and Asteroid Bombardment. Reading, Mass.: Addison-Wesley, 1996.

McGrayne, Sharon Bertsch, Prometheans in the Lab: Chemistry and the Making of the Modern World. London: McGraw-Hill, 2002.

McGuire, Bill, A Guide to the End of the World: Everything You Never Wanted to Know. Oxford: Oxford University Press, 2002.

McSween, Harry Y., Jr, Stardust to Planets: A Geological Tour of the Solar System. New York: St Martin's Press, 1993.

Officer, Charles, and Jake Page, Tales of the Earth: Paroxysms and Perturbations of the Blue Planet. New York: Oxford University Press, 1993.

Oldstone, Michael B. A., Viruses, Plagues and History. New York: Oxford University Press, 1998.

Overbye, Dennis, Lonely Hearts of the Cosmos: The Scientific Quest for the Secret of the Universe. London: Macmillan, 1991.

Ozima, Minoru, The Earth: It's Birth and Growth. Cambridge: Cambridge University Press, 1981.

Parker, Ronald B., Inscrutable Earth: Explorations in the Science of Earth. New York: Charles Scribner's Sons, 1984.

Rees, Martin, Just Six Numbers: The Deep Forces that Shape the Universe. London: Phoenix / Orion, 2000.

Snow, C. P., The Physicists. London: House of Stratus, 1979.

Strathern, Paul, Mendeleyev's Dream: The Quest for the Elements. London: Penguin Books, 2001.

Thomas, Keith, Man and the Natural World: Changing Attitudes in England, 1500-1800. London: Penguin Books, 1984.

Trefil, James, The Unexpected Vista: A Physicist's View of Nature. New York: Charles Scribner's Sons, 1987.

Westfall, Richard S., Never At Rest: A Biography of Isaac Newton, New York: Cambridge University Press. 1980

White, Michael, Isaac Newton: The Last Sorcerer. London: Fourth Dimension, 1997.

Answers to Newtons Law of cooling activities

Q1   T(1200 s) = 293.0 K + 13.224 K

T(1200 s) ≈ 306.224 K

After 20 minutes, the soup's temperature is

306.224K.

$$k = -\frac{1}{10.0\ s}(-3.002)$$

Q2

$$k = \frac{3.002}{10.0\ s}$$

$k = 0.300\ 1/s$

The cooling constant for the iron rod in water is $k = 0.300\ 1/s$.

Answers to Alchemy activity

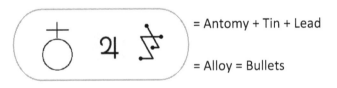

= Antomy + Tin + Lead

= Alloy = Bullets

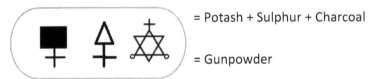

= Potash + Sulphur + Charcoal

= Gunpowder

Printed in Great Britain
by Amazon

27400105R00117